Graph Representation Learning

Synthesis Lectures on Artifical Intelligence and Machine Learning

Editors
Ronald Brachman, *Jacobs Technion-Cornell Institute at Cornell Tech*
Francesca Rossi, *IBM Research AI*
Peter Stone, *University of Texas at Austin*

Essentials of Game Theory: A Concise Multidisciplinary Introduction
Kevin Leyton-Brown and Yoav Shoham
2008

A Concise Introduction to Multiagent Systems and Distributed Artificial Intelligence
Nikos Vlassis
2007

Intelligent Autonomous Robotics: A Robot Soccer Case Study
Peter Stone
2007

Graph Representation Learning

William L. Hamilton

ISBN: 978-3-031-00460-5   paperback
ISBN: 978-3-031-01588-5   ebook
ISBN: 978-3-031-00033-1   hardcover

DOI 10.1007/978-3-031-01588-5

A Publication in the Springer series
SYNTHESIS LECTURES ON ARTIFICAL INTELLIGENCE AND MACHINE LEARNING

Lecture #46
Series Editors: Ronald Brachman, *Jacobs Technion–Cornell Institute at Cornell Tech*
                Francesca Rossi, *IBM Research AI*
                Peter Stone, *University of Texas at Austin*
Series ISSN
Synthesis Lectures on Artifical Intelligence and Machine Learning
Print 1939-4608   Electronic 1939-4616

Graph Representation Learning

William L. Hamilton
McGill University and Mila-Quebec Artificial Intelligence Institute

SYNTHESIS LECTURES ON ARTIFICAL INTELLIGENCE AND MACHINE LEARNING #46

ABSTRACT

Graph-structured data is ubiquitous throughout the natural and social sciences, from telecommunication networks to quantum chemistry. Building relational inductive biases into deep learning architectures is crucial for creating systems that can learn, reason, and generalize from this kind of data. Recent years have seen a surge in research on graph representation learning, including techniques for deep graph embeddings, generalizations of convolutional neural networks to graph-structured data, and neural message-passing approaches inspired by belief propagation. These advances in graph representation learning have led to new state-of-the-art results in numerous domains, including chemical synthesis, 3D vision, recommender systems, question answering, and social network analysis.

This book provides a synthesis and overview of graph representation learning. It begins with a discussion of the goals of graph representation learning as well as key methodological foundations in graph theory and network analysis. Following this, the book introduces and reviews methods for learning node embeddings, including random-walk-based methods and applications to knowledge graphs. It then provides a technical synthesis and introduction to the highly successful graph neural network (GNN) formalism, which has become a dominant and fast-growing paradigm for deep learning with graph data. The book concludes with a synthesis of recent advancements in deep generative models for graphs—a nascent but quickly growing subset of graph representation learning.

KEYWORDS

graph neural networks, graph embeddings, node embeddings, deep learning, relational data, knowledge graphs, social networks, network analysis, graph signal processing, graph convolutions, spectral graph theory, geometric deep learning

Contents

Preface

The field of graph representation learning has grown at an incredible—and sometimes unwieldy—pace over the past seven years. I first encountered this area as a graduate student in 2013, during the time when many researchers began investigating deep learning methods for "embedding" graph-structured data. In the years since 2013, the field of graph representation learning has witnessed a truly impressive rise and expansion—from the development of the standard graph neural network paradigm to the nascent work on deep generative models of graph-structured data. The field has transformed from a small subset of researchers working on a relatively niche topic to one of the fastest growing sub-areas of deep learning.

However, as the field has grown, our understanding of the methods and theories underlying graph representation learning has also stretched backwards through time. We can now view the popular "node embedding" methods as well-understood extensions of classic work on dimensionality reduction. We now have an understanding and appreciation for how graph neural networks evolved—somewhat independently—from historically rich lines of work on spectral graph theory, harmonic analysis, variational inference, and the theory of graph isomorphism. This book is my attempt to synthesize and summarize these methodological threads in a practical way. My hope is to introduce the reader to the current practice of the field while also connecting this practice to broader lines of historical research in machine learning and beyond.

Intended audience This book is intended for a graduate-level researcher in machine learning or an advanced undergraduate student. The discussions of graph-structured data and graph properties are relatively self-contained. However, the book does assume a background in machine learning and a familiarity with modern deep learning methods (e.g., convolutional and recurrent neural networks). Generally, the book assumes a level of machine learning and deep learning knowledge that one would obtain from a textbook such as Goodfellow et al. [2016]'s *Deep Learning* book.

William L. Hamilton
September 2020

Acknowledgments

Over the past several years, I have had the good fortune to work with many outstanding collaborators on topics related to graph representation learning—many of whom have made seminal contributions to this nascent field. I am deeply indebted to all these collaborators and friends: my colleagues at Stanford, McGill, University of Toronto, and elsewhere; my graduate students at McGill—who taught me more than anyone else the value of pedagogical writing; and my Ph.D. advisors—Dan Jurafsky and Jure Leskovec—who encouraged and seeded this path for my research.

I also owe a great debt of gratitude to the students of my Winter 2020 graduate seminar at McGill University. These students were the early "beta testers" of this material, and this book would not exist without their feedback and encouragement. In a similar vein, the exceptionally detailed feedback provided by Petar Velčković, as well as comments by Mariá C. V. Nascimento and Jian Tang, were invaluable during revisions of the manuscript.

No book is written in a vacuum. This book is the culmination of years of collaborations with many outstanding colleagues—not to mention months of support from my wife and partner, Amy. It is safe to say that this book could not have been written without their support. Though, of course, any errors are mine alone.

William L. Hamilton
September 2020

CHAPTER 1

Introduction

Graphs are a ubiquitous data structure and a universal language for describing complex systems. In the most general view, a graph is simply a collection of objects (i.e., nodes), along with a set of interactions (i.e., edges) between pairs of these objects. For example, to encode a social network as a graph we might use nodes to represent individuals and use edges to represent that two individuals are friends (Figure 1.1). In the biological domain we could use the nodes in a graph to represent proteins, and use the edges to represent various biological interactions, such as kinetic interactions between proteins.

The power of the graph formalism lies both in its focus on *relationships between points* (rather than the properties of individual points), as well as in its generality. The same graph formalism can be used to represent social networks, interactions between drugs and proteins, the interactions between atoms in a molecule, or the connections between terminals in a telecommunications network—to name just a few examples.

Graphs do more than just provide an elegant theoretical framework, however. They offer a mathematical foundation that we can build upon to analyze, understand, and learn from real-world complex systems. In the last 25 years, there has been a dramatic increase in the quantity and quality of graph-structured data that is available to researchers. With the advent of large-scale social networking platforms, massive scientific initiatives to model the interactome, food webs, databases of molecule graph structures, and billions of interconnected web-enabled devices, there is no shortage of meaningful graph data for researchers to analyze. The challenge is unlocking the potential of this data.

This book is about how we can use *machine learning* to tackle this challenge. Of course, machine learning is not the only possible way to analyze graph data.[1] However, given the ever-increasing scale and complexity of the graph datasets that we seek to analyze, it is clear that machine learning will play an important role in advancing our ability to model, analyze, and understand graph data.

1.1 WHAT IS A GRAPH?

Before we discuss machine learning on graphs, it is necessary to give a bit more formal description of what exactly we mean by "graph data." Formally, a graph $\mathcal{G} = (\mathcal{V}, \mathcal{E})$ is defined by a set of nodes $\mathcal{V}$ and a set of edges $\mathcal{E}$ between these nodes. We denote an edge going from node $u \in \mathcal{V}$

---

[1]The field of *network analysis* independent of machine learning is the subject of entire textbooks and will not be covered in detail here [Newman, 2018].

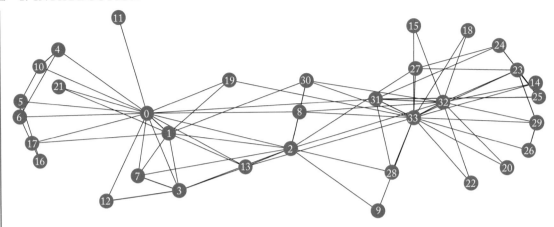

Figure 1.1: The famous *Zachary Karate Club Network* represents the friendship relationships between members of a karate club studied by Wayne W. Zachary from 1970–1972. An edge connects two individuals if they socialized outside of the club. During Zachary's study, the club split into two factions—centered around nodes 0 and 33—and Zachary was able to correctly predict which nodes would fall into each faction based on the graph structure [Zachary, 1977].

to node $v \in \mathcal{V}$ as $(u, v) \in \mathcal{E}$. In many cases we will be concerned only with *simple graphs*, where there is at most one edge between each pair of nodes, no edges between a node and itself, and where the edges are all undirected, i.e., $(u, v) \in \mathcal{E} \leftrightarrow (v, u) \in \mathcal{E}$.

A convenient way to represent graphs is through an *adjacency matrix* $\mathbf{A} \in \mathbb{R}^{|\mathcal{V}| \times |\mathcal{V}|}$. To represent a graph with an adjacency matrix, we order the nodes in the graph so that every node indexes a particular row and column in the adjacency matrix. We can then represent the presence of edges as entries in this matrix: $\mathbf{A}[u, v] = 1$ if $(u, v) \in \mathcal{E}$ and $\mathbf{A}[u, v] = 0$, otherwise. If the graph contains only undirected edges then $\mathbf{A}$ will be a symmetric matrix, but if the graph is *directed* (i.e., edge direction matters) then $\mathbf{A}$ will not necessarily be symmetric. Some graphs can also have *weighted* edges, where the entries in the adjacency matrix are arbitrary real-values rather than $\{0, 1\}$. For instance, a weighted edge in a protein-protein interaction graph might indicated the strength of the association between two proteins.

1.1.1 MULTI-RELATIONAL GRAPHS

Beyond the distinction between undirected, directed, and weighted edges, we will also consider graphs that have different *types* of edges. For instance, in graphs representing drug-drug interactions, we might want different edges to correspond to different side effects that can occur when you take a pair of drugs at the same time. In these cases we can extend the edge notation to include an edge or relation type τ, e.g., $(u, \tau, v) \in \mathcal{E}$, and we can define one adjacency matrix $\mathbf{A}_\tau$ per edge type. We call such graphs *multi-relational*, and the entire graph can be summarized by an adjacency tensor $\mathcal{A} \in \mathbb{R}^{|\mathcal{V}| \times |\mathcal{R}| \times |\mathcal{V}|}$, where $\mathcal{R}$ is the set of relations. Two important subsets of multi-relational graphs are often known as *heterogeneous* and *multiplex* graphs.

Heterogeneous graphs   In heterogeneous graphs, nodes are also imbued with *types*, meaning that we can partition the set of nodes into disjoint sets $\mathcal{V} = \mathcal{V}_1 \cup \mathcal{V}_2 \cup ... \cup \mathcal{V}_k$ where $\mathcal{V}_i \cap \mathcal{V}_j = \emptyset, \forall i \neq j$. Edges in heterogeneous graphs generally satisfy constraints according to the node types, most commonly the constraint that certain edges only connect nodes of certain types, i.e., $(u, \tau_i, v) \in \mathcal{E} \rightarrow u \in \mathcal{V}_j, v \in \mathcal{V}_k$. For example, in a heterogeneous biomedical graph, there might be one type of node representing proteins, one type of representing drugs, and one type representing diseases. Edges representing "treatments" would only occur between drug nodes and disease nodes. Similarly, edges representing "polypharmacy side-effects" would only occur between two drug nodes. *Multipartite* graphs are a well-known special case of heterogeneous graphs, where edges can only connect nodes that have different types, i.e., $(u, \tau_i, v) \in \mathcal{E} \rightarrow u \in \mathcal{V}_j, v \in \mathcal{V}_k \wedge j \neq k$.

Multiplex graphs   In multiplex graphs we assume that the graph can be decomposed in a set of *k layers*. Every node is assumed to belong to every layer, and each layer corresponds to a unique relation, representing the *intra-layer* edge type for that layer. We also assume that *inter-layer* edges types can exist, which connect the same node across layers. Multiplex graphs are best understood via examples. For instance, in a multiplex transportation network, each node might represent a city and each layer might represent a different mode of transportation (e.g., air travel or train travel). Intra-layer edges would then represent cities that are connected by different modes of transportation, while inter-layer edges represent the possibility of switching modes of transportation within a particular city.

1.1.2   FEATURE INFORMATION

Last, in many cases we also have *attribute* or *feature* information associated with a graph (e.g., a profile picture associated with a user in a social network). Most often these are node-level attributes that we represent using a real-valued matrix $\mathbf{X} \in \mathbb{R}^{|V| \times m}$, where we assume that the ordering of the nodes is consistent with the ordering in the adjacency matrix. In heterogeneous graphs we generally assume that each different type of node has its own distinct type of attributes. In rare cases we will also consider graphs that have real-valued edge features in addition to discrete edge types, and in some cases we even associate real-valued features with entire graphs.

> **Graph or network?**   We use the term "graph" in this book, but you will see many other resources use the term "network" to describe the same kind of data. In some places, we will use both terms (e.g., for social or biological networks). So which term is correct? In many ways, this terminological difference is a historical and cultural one: the term "graph" appears to be more prevalent in machine learning community,[a] but "network" has historically been popular in the data mining and (unsurprisingly) network science communities. We use both

terms in this book, but we also make a distinction between the usage of these terms. We use the term *graph* to describe the abstract data structure that is the focus of this book, but we will also often use the term *network* to describe specific, real-world instantiations of this data structure (e.g., social networks). This terminological distinction is fitting with their current popular usages of these terms. *Network analysis* is generally concerned with the properties of real-world data, whereas *graph theory* is concerned with the theoretical properties of the mathematical graph abstraction.

[a]Perhaps in some part due to the terminological clash with "neural networks."

1.2   MACHINE LEARNING ON GRAPHS

Machine learning is inherently a problem-driven discipline. We seek to build models that can learn from data in order to solve particular tasks, and machine learning models are often categorized according to the type of task they seek to solve: Is it a *supervised* task, where the goal is to predict a target output given an input datapoint? Is it an *unsupervised* task, where the goal is to infer patterns, such as clusters of points, in the data?

Machine learning with graphs is no different, but the usual categories of supervised and unsupervised are not necessarily the most informative or useful when it comes to graphs. In this section we provide a brief overview of the most important and well-studied machine learning tasks on graph data. As we will see, "supervised" problems are popular with graph data, but machine learning problems on graphs often blur the boundaries between the traditional machine learning categories.

1.2.1   NODE CLASSIFICATION

Suppose we are given a large social network dataset with millions of users, but we know that a significant number of these users are actually bots. Identifying these bots could be important for many reasons: a company might not want to advertise to bots or bots may actually be in violation of the social network's terms of service. Manually examining every user to determine if they are a bot would be prohibitively expensive, so ideally we would like to have a model that could classify users as a bot (or not) given only a small number of manually labeled examples.

This is a classic example of *node classification*, where the goal is to predict the label y_u—which could be a type, category, or attribute—associated with all the nodes $u \in \mathcal{V}$, when we are only given the true labels on a *training set* of nodes $\mathcal{V}_{\text{train}} \subset \mathcal{V}$. Node classification is perhaps the most popular machine learning task on graph data, especially in recent years. Examples of node classification beyond social networks include classifying the function of proteins in the interactome [Hamilton et al., 2017b] and classifying the topic of documents based on hyperlink or citation graphs [Kipf and Welling, 2016a]. Often, we assume that we have label information only for a very small subset of the nodes in a single graph (e.g., classifying bots in a social network from a small set of manually labeled examples). However, there are also instances of node classification that involve many labeled nodes and/or that require generalization across disconnected graphs (e.g., classifying the function of proteins in the interactomes of different species).

At first glance, node classification appears to be a straightforward variation of standard supervised classification, but there are in fact important differences. The most important difference is that the nodes in a graph are not *independent and identically distributed (i.i.d.)*. Usually, when we build supervised machine learning models we assume that each datapoint is statistically *independent* from all the other datapoints; otherwise, we might need to model the dependencies between all our input points. We also assume that the datapoints are *identically distributed*; otherwise, we have no way of guaranteeing that our model will generalize to new datapoints. Node classification completely breaks this i.i.d. assumption. Rather than modeling a set of i.i.d. datapoints, we are instead modeling an interconnected set of nodes.

In fact, the key insight behind many of the most successful node classification approaches is to explicitly leverage the connections between nodes. One particularly popular idea is to exploit *homophily*, which is the tendency for nodes to share attributes with their neighbors in the graph [McPherson et al., 2001]. For example, people tend to form friendships with others who share the same interests or demographics. Based on the notion of homophily we can build machine learning models that try to assign similar labels to neighboring nodes in a graph [Zhou et al., 2004]. Beyond homophily there are also concepts such as *structural equivalence* [Donnat et al., 2018], which is the idea that nodes with similar local neighborhood structures will have similar labels, as well as *heterophily*, which presumes that nodes will be preferentially connected to nodes with different labels.[2] When we build node classification models we want to exploit these concepts and model the relationships between nodes, rather than simply treating nodes as independent datapoints.

Supervised or semi-supervised? Due to the atypical nature of node classification, researchers often refer to it as *semi-supervised* [Yang et al., 2016]. This terminology is used because when we are training node classification models, we usually have access to the full graph, including all the unlabeled (e.g., test) nodes. The only thing we are missing is the labels of test nodes. However, we can still use information about the test nodes (e.g., knowledge of their neighborhood in the graph) to improve our model during training. This is different from the usual supervised setting, in which unlabeled datapoints are completely unobserved during training.

The general term used for models that combine labeled and unlabeled data during training is semi-supervised learning, so it is understandable that this term is often used in reference to node classification tasks. It is important to note, however, that standard formulations of semi-supervised learning still require the i.i.d. assumption, which does not

---

[2]For example, gender is an attribute that exhibits heterophily in many social networks.

hold for node classification. Machine learning tasks on graphs do not easily fit our standard categories!

1.2.2   RELATION PREDICTION

Node classification is useful for inferring information about a node based on its relationship with other nodes in the graph. But what about cases where we are missing this relationship information? What if we know only some of protein-protein interactions that are present in a given cell, but we want to make a good guess about the interactions we are missing? Can we use machine learning to infer the edges between nodes in a graph?

This task goes by many names, such as link prediction, graph completion, and relational inference, depending on the specific application domain. We will simply call it *relation prediction* here. Along with node classification, it is one of the more popular machine learning tasks with graph data and has countless real-world applications: recommending content to users in social platforms [Ying et al., 2018a], predicting drug side-effects [Zitnik et al., 2018], or inferring new facts in a relational databases [Bordes et al., 2013]—all of these tasks can be viewed as special cases of relation prediction.

The standard setup for relation prediction is that we are given a set of nodes $\mathcal{V}$ and an incomplete set of edges between these nodes $\mathcal{E}_{\text{train}} \subset \mathcal{E}$. Our goal is to use this partial information to infer the missing edges $\mathcal{E} \setminus \mathcal{E}_{\text{train}}$. The complexity of this task is highly dependent on the type of graph data we are examining. For instance, in simple graphs, such as social networks that only encode "friendship" relations, there are simple heuristics based on how many neighbors two nodes share that can achieve strong performance [Lü and Zhou, 2011]. On the other hand, in more complex multi-relational graph datasets, such as biomedical knowledge graphs that encode hundreds of different biological interactions, relation prediction can require complex reasoning and inference strategies [Nickel et al., 2016]. Like node classification, relation prediction blurs the boundaries of traditional machine learning categories—often being referred to as both supervised and unsupervised—and it requires inductive biases that are specific to the graph domain. In addition, like node classification, there are many variants of relation prediction, including settings where the predictions are made over a single, fixed graph [Lü and Zhou, 2011], as well as settings where relations must be predicted across multiple disjoint graphs [Teru et al., 2020].

1.2.3   CLUSTERING AND COMMUNITY DETECTION

Both node classification and relation prediction require inferring *missing* information about graph data, and in many ways, those two tasks are the graph analogs of supervised learning. *Community detection*, on the other hand, is the graph analog of unsupervised clustering.

Suppose we have access to all the citation information in Google Scholar, and we make a *collaboration graph* that connects two researchers if they have co-authored a paper together. If we were to examine this network, would we expect to find a dense "hairball" where everyone is equally likely to collaborate with everyone else? It is more likely that the graph would segregate into different *clusters* of nodes, grouped together by research area, institution, or other demographic factors. In other words, we would expect this network—like many real-world networks—to exhibit a *community* structure, where nodes are much more likely to form edges with nodes that belong to the same community.

This is the general intuition underlying the task of community detection. The challenge of community detection is to infer latent community structures given only the input graph $\mathcal{G} = (\mathcal{V}, \mathcal{E})$. The many real-world applications of community detection include uncovering functional modules in genetic interaction networks [Agrawal et al., 2018] and uncovering fraudulent groups of users in financial transaction networks [Pandit et al., 2007].

1.2.4   GRAPH CLASSIFICATION, REGRESSION, AND CLUSTERING

The final class of popular machine learning applications on graph data involve classification, regression, or clustering problems over entire graphs. For instance, given a graph representing the structure of a molecule, we might want to build a regression model that could predict that molecule's toxicity or solubility [Gilmer et al., 2017]. Or, we might want to build a classification model to detect whether a computer program is malicious by analyzing a graph-based representation of its syntax and data flow [Li et al., 2019]. In these *graph classification or regression* applications, we seek to learn over graph data, but instead of making predictions over the individual components of a single graph (i.e., the nodes or the edges), we are instead given a dataset of *multiple different graphs* and our goal is to make independent predictions specific to each graph. In the related task of *graph clustering*, the goal is to learn an unsupervised measure of similarity between pairs of graphs.

Of all the machine learning tasks on graphs, graph regression and classification are perhaps the most straightforward analogs of standard supervised learning. Each graph is an i.i.d. datapoint associated with a label, and the goal is to use a labeled set of training points to learn a mapping from datapoints (i.e., graphs) to labels. In a similar way, graph clustering is the straightforward extension of unsupervised clustering for graph data. The challenge in these graph-level tasks, however, is how to define useful features that take into account the relational structure within each datapoint.

CHAPTER 2

Background and Traditional Approaches

Before we introduce the concepts of graph representation learning and deep learning on graphs, it is necessary to give some methodological background and context. What kinds of methods were used for machine learning on graphs prior to the advent of modern deep learning approaches? In this chapter, we will provide a very brief and focused tour of traditional learning approaches over graphs, providing pointers and references to more thorough treatments of these methodological approaches along the way. This background chapter will also serve to introduce key concepts from graph analysis that will form the foundation for later chapters.

Our tour will be roughly aligned with the different kinds of learning tasks on graphs. We will begin with a discussion of basic graph statistics, kernel methods, and their use for node and graph classification tasks. Following this, we will introduce and discuss various approaches for measuring the overlap between node neighborhoods, which form the basis of strong heuristics for relation prediction. Finally, we will close this background section with a brief introduction of spectral clustering using graph Laplacians. Spectral clustering is one of the most well-studied algorithms for clustering or community detection on graphs, and our discussion of this technique will also introduce key mathematical concepts that will re-occur throughout this book.

2.1 GRAPH STATISTICS AND KERNEL METHODS

Traditional approaches to classification using graph data follow the standard machine learning paradigm that was popular prior to the advent of deep learning. We begin by extracting some statistics or features—based on heuristic functions or domain knowledge—and then use these features as input to a standard machine learning classifier (e.g., logistic regression). In this section, we will first introduce some important node-level features and statistics, and we will follow this by a discussion of how these node-level statistics can be generalized to graph-level statistics and extended to design kernel methods over graphs. Our goal will be to introduce various heuristic statistics and graph properties, which are often used as features in traditional machine learning pipelines applied to graphs.

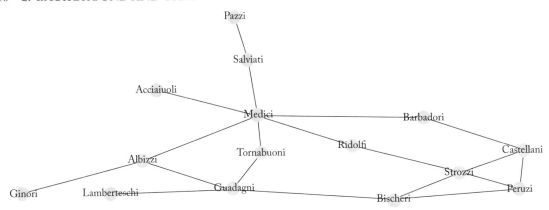

Figure 2.1: A visualization of the marriages between various different prominent families in 15th-century Florence [Padgett and Ansell, 1993].

2.1.1   NODE-LEVEL STATISTICS AND FEATURES

Following Jackson [2010], we will motivate our discussion of node-level statistics and features with a simple (but famous) social network: the network of 15th-century Florentine marriages (Figure 2.1). This social network is well known due to the work of Padgett and Ansell [1993], which used this network to illustrate the rise in power of the Medici family (depicted near the center) who came to dominate Florentine politics. Political marriages were an important way to consolidate power during the era of the Medicis, so this network of marriage connections encodes a great deal about the political structure of this time.

    For our purposes, we will consider this network and the rise of the Medici from a machine learning perspective and ask the question: What features or statistics could a machine learning model use to predict the Medici's rise? In other words, what properties or statistics of the Medici node distinguish it from the rest of the graph? And, more generally, what are useful properties and statistics that we can use to characterize the nodes in this graph?

    In principle, the properties and statistics we discuss below could be used as features in a node classification model (e.g., as input to a logistic regression model). Of course, we would not be able to realistically train a machine learning model on a graph as small as the Florentine marriage network. However, it is still illustrative to consider the kinds of features that could be used to differentiate the nodes in such a real-world network, and the properties we discuss are generally useful across a wide variety of node classification tasks.

Node degree.   The most obvious and straightforward node feature to examine is *degree*, which is usually denoted d_u for a node $u \in \mathcal{V}$ and simply counts the number of edges incident to a node:

$$d_u = \sum_{v \in V} \mathbf{A}[u, v].$$ (2.1)

Note that in cases of directed and weighted graphs, one can differentiate between different notions of degree—e.g., corresponding to outgoing edges or incoming edges by summing over rows or columns in Equation (2.1). In general, the degree of a node is an essential statistic to consider, and it is often one of the most informative features in traditional machine learning models applied to node-level tasks.

In the case of our illustrative Florentine marriages graph, we can see that degree is indeed a good feature to distinguish the Medici family, as they have the highest degree in the graph. However, their degree only outmatches the two closest families—the Strozzi and the Guadagni—by a ratio of 3–2. Are there perhaps additional or more discriminative features that can help to distinguish the Medici family from the rest of the graph?

Node Centrality

Node degree simply measures how many neighbors a node has, but this is not necessarily sufficient to measure the *importance* of a node in a graph. In many cases—such as our example graph of Florentine marriages—we can benefit from additional and more powerful measures of node importance. To obtain a more powerful measure of importance, we can consider various measures of what is known as node *centrality*, which can form useful features in a wide variety of node classification tasks.

One popular and important measure of centrality is the so-called *eigenvector centrality*. Whereas degree simply measures how many neighbors each node has, eigenvector centrality also takes into account how important a node's neighbors are. In particular, we define a node's eigenvector centrality e_u via a recurrence relation in which the node's centrality is proportional to the average centrality of its neighbors:

$$e_u = \frac{1}{\lambda} \sum_{v \in V} \mathbf{A}[u, v] e_v \ \forall u \in \mathcal{V},$$ (2.2)

where λ is a constant. Rewriting this equation in vector notation with $\mathbf{e}$ as the vector of node centralities, we can see that this recurrence defines the standard eigenvector equation for the adjacency matrix:

$$\lambda \mathbf{e} = \mathbf{A} \mathbf{e}.$$ (2.3)

In other words, the centrality measure that satisfies the recurrence in Equation (2.2) corresponds to an eigenvector of the adjacency matrix. Assuming that we require positive centrality values, we can apply the Perron-Frobenius Theorem[1] to further determine that the vector of centrality

---

[1]The Perron-Frobenius Theorem is a fundamental result in linear algebra, proved independently by Oskar Perron and Georg Frobenius [Meyer, 2000]. The full theorem has many implications, but for our purposes the key assertion in the theorem is that any irreducible square matrix has a unique largest real eigenvalue, which is the only eigenvalue whose corresponding eigenvector can be chosen to have strictly positive components.

values $\mathbf{e}$ is given by the eigenvector corresponding to the largest eigenvalue of $\mathbf{A}$ [Newman, 2016].

One view of eigenvector centrality is that it ranks the likelihood that a node is visited on a random walk of infinite length on the graph. This view can be illustrated by considering the use of power iteration to obtain the eigenvector centrality values. That is, since λ is the leading eigenvector of $\mathbf{A}$, we can compute $\mathbf{e}$ using power iteration via[2]

$$\mathbf{e}^{(t+1)} = \mathbf{A}\mathbf{e}^{(t)}. \tag{2.4}$$

If we start off this power iteration with the vector $\mathbf{e}^{(0)} = (1, 1, ..., 1)^{\top}$, then we can see that after the first iteration $\mathbf{e}^{(1)}$ will contain the degrees of all the nodes. In general, at iteration $t \geq 1$, $\mathbf{e}^{(t)}$ will contain the number of length-t paths arriving at each node. Thus, by iterating this process indefinitely we obtain a score that is proportional to the number of times a node is visited on paths of infinite length. This connection between node importance, random walks, and the spectrum of the graph adjacency matrix will return often throughout the ensuing sections and chapters of this book.

Returning to our example of the Florentine marriage network, if we compute the eigenvector centrality values on this graph, we again see that the Medici family is the most influential, with a normalized value of 0.43 compared to the next-highest value of 0.36. There are, of course, other measures of centrality that we could use to characterize the nodes in this graph—some of which are even more discerning with respect to the Medici family's influence. These include *betweeness centrality*—which measures how often a node lies on the shortest path between two other nodes—as well as *closeness centrality*—which measures the average shortest path length between a node and all other nodes. These measures and many more are reviewed in detail by Newman [2018].

The Clustering Coefficient

Measures of importance, such as degree and centrality, are clearly useful for distinguishing the prominent Medici family from the rest of the Florentine marriage network. But what about features that are useful for distinguishing between the other nodes in the graph? For example, the Peruzzi and Guadagni nodes in the graph have very similar degree (3 vs. 4) and similar eigenvector centralities (0.28 vs. 0.29). However, looking at the graph in Figure 2.1, there is a clear difference between these two families. Whereas the Peruzzi family is in the midst of a relatively tight-knit cluster of families, the Guadagni family occurs in a more "star-like" role.

This important structural distinction can be measured using variations of the *clustering coefficient*, which measures the proportion of closed triangles in a node's local neighborhood. The popular *local variant* of the clustering coefficient is computed as follows [Watts and Strogatz,

---

[2]Note that we have ignored the normalization in the power iteration computation for simplicity, as this does not change the main result.

1998]:

$$c_u = \frac{|(v_1, v_2) \in \mathcal{E} \; : \; v_1, v_2 \in \mathcal{N}(u)|}{\binom{d_u}{2}}. \tag{2.5}$$

The numerator in this equation counts the number of edges between neighbors of node u (where we use $\mathcal{N}(u) = \{v \in \mathcal{V} \; : \; (u, v) \in \mathcal{E}\}$ to denote the node neighborhood). The denominator calculates how many pairs of nodes there are in u's neighborhood.

The clustering coefficient takes its name from the fact that it measures how tightly clustered a node's neighborhood is. A clustering coefficient of 1 would imply that all of u's neighbors are also neighbors of each other. In our Florentine marriage graph, we can see that some nodes are highly clustered—e.g., the Peruzzi node has a clustering coefficient of 0.66—while other nodes such as the Guadagni node have clustering coefficients of 0. As with centrality, there are numerous variations of the clustering coefficient (e.g., to account for directed graphs), which are also reviewed in detail by Newman [2018]. An interesting and important property of real-world networks throughout the social and biological sciences is that they tend to have far higher clustering coefficients than one would expect if edges were sampled randomly [Watts and Strogatz, 1998].

Closed Triangles, Ego Graphs, and Motifs

An alternative way of viewing the clustering coefficient—rather than as a measure of local clustering—is that it counts the number of closed triangles within each node's local neighborhood. In more precise terms, the clustering coefficient is related to the ratio between the actual number of triangles and the total possible number of triangles within a node's *ego graph*, i.e., the subgraph containing that node, its neighbors, and all the edges between nodes in its neighborhood.

This idea can be generalized to the notion of counting arbitrary *motifs* or *graphlets* within a node's ego graph. That is, rather than just counting triangles, we could consider more complex structures, such as cycles of particular length, and we could characterize nodes by counts of how often these different motifs occur in their ego graph. Indeed, by examining a node's ego graph in this way, we can essentially transform the task of computing node-level statistics and features to a graph-level task. Thus, we will now turn our attention to this graph-level problem.

2.1.2 GRAPH-LEVEL FEATURES AND GRAPH KERNELS

So far we have discussed various statistics and properties at the node level, which could be used as features for node-level classification tasks. However, what if our goal is to do graph-level classification? For example, suppose we are given a dataset of graphs representing molecules and our goal is to classify the solubility of these molecules based on their graph structure. How would we do this? In this section, we will briefly survey approaches to extracting graph-level features for such tasks.

Many of the methods we survey here fall under the general classification of *graph kernel methods*, which are approaches to designing features for graphs or implicit kernel functions that can be used in machine learning models. We will touch upon only a small fraction of the approaches within this large area, and we will focus on methods that extract explicit feature representations, rather than approaches that define implicit kernels (i.e., similarity measures) between graphs. We point the interested reader to Kriege et al. [2020] and Vishwanathan et al. [2010] for detailed surveys of this area.

Bag of Nodes

The simplest approach to defining a graph-level feature is to just aggregate node-level statistics. For example, one can compute histograms or other summary statistics based on the degrees, centralities, and clustering coefficients of the nodes in the graph. This aggregated information can then be used as a graph-level representation. The downside to this approach is that it is entirely based upon local node-level information and can miss important global properties in the graph.

The Weisfieler–Lehman Kernel

One way to improve the basic bag of nodes approach is using a strategy of *iterative neighborhood aggregation*. The idea with these approaches is to extract node-level features that contain more information than just their local ego graph, and then to aggregate these richer features into a graph-level representation.

Perhaps the most important and well known of these strategies is the Weisfieler–Lehman (WL) algorithm and kernel [Shervashidze et al., 2011, Weisfeiler and Lehman, 1968]. The basic idea behind the WL algorithm is the following:

1. First, we assign an initial label $l^{(0)}(v)$ to each node. In most graphs, this label is simply the degree, i.e., $l^{(0)}(v) = d_v \ \forall v \in V$.

2. Next, we iteratively assign a new label to each node by hashing the multi-set of the current labels within the node's neighborhood:

$$l^{(i)}(v) = \text{HASH}(\{\{l^{(i-1)}(u) \ \forall u \in \mathcal{N}(v)\}\}), \tag{2.6}$$

   where the double-braces are used to denote a multi-set and the HASH function maps each unique multi-set to a unique new label.

3. After running K iterations of re-labeling (i.e., Step 2), we now have a label $l^{(K)}(v)$ for each node that summarizes the structure of its K-hop neighborhood. We can then compute histograms or other summary statistics over these labels as a feature representation for the graph. In other words, the WL kernel is computed by measuring the difference between the resultant label sets for two graphs.

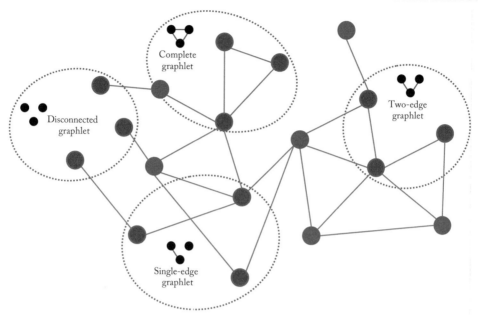

Figure 2.2: The four different size-3 graphlets that can occur in a simple graph.

The WL kernel is popular, well studied, and known to have important theoretical properties. For example, one popular way to approximate graph isomorphism is to check whether or not two graphs have the same label set after K rounds of the WL algorithm, and this approach is known to solve the isomorphism problem for a broad set of graphs [Shervashidze et al., 2011].

Graphlets and Path-Based Methods

Finally, just as in our discussion of node-level features, one valid and powerful strategy for defining features over graphs is to simply count the occurrence of different small subgraph structures, usually called *graphlets* in this context. Formally, the graphlet kernel involves enumerating all possible graph structures of a particular size and counting how many times they occur in the full graph. (Figure 2.2 illustrates the various graphlets of size 3.) The challenge with this approach is that counting these graphlets is a combinatorially difficult problem, though numerous approximations have been proposed [Shervashidze and Borgwardt, 2009].

An alternative to enumerating all possible graphlets is to use *path-based methods*. In these approaches, rather than enumerating graphlets, one simply examines the different kinds of *paths* that occur in the graph. For example, the random walk kernel proposed by Kashima et al. [2003] involves running random walks over the graph and then counting the occurrence of different degree sequences,[3] while the shortest-path kernel of Borgwardt and Kriegel [2005] involves a

---

[3]Other node labels can also be used.

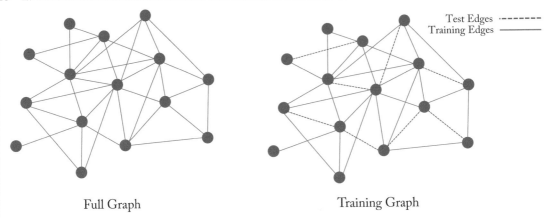

Figure 2.3: An illustration of a full graph and a subsampled graph used for training. The dotted edges in the training graph are removed when training a model or computing the overlap statistics. The model is evaluated based on its ability to predict the existence of these held-out *test edges*.

similar idea but uses only the shortest-paths between nodes (rather than random walks). As we will see in Chapter 3 of this book, this idea of characterizing graphs based on walks and paths is a powerful one, as it can extract rich structural information while avoiding many of the combinatorial pitfalls of graph data.

2.2   NEIGHBORHOOD OVERLAP DETECTION

In the last section we covered various approaches to extract features or statistics about individual nodes or entire graphs. These node and graph-level statistics are useful for many classification tasks. However, they are limited in that they do not quantify the *relationships* between nodes. For instance, the statistics discussed in the last section are not very useful for the task of relation prediction, where our goal is to predict the existence of an edge between two nodes (Figure 2.3).

In this section we will consider various statistical measures of neighborhood overlap between pairs of nodes, which quantify the extent to which a pair of nodes are related. For example, the simplest neighborhood overlap measure just counts the number of neighbors that two nodes share:

$$\mathbf{S}[u, v] = |\mathcal{N}(u) \cap \mathcal{N}(v)|, \tag{2.7}$$

where we use $\mathbf{S}[u, v]$ to denote the value quantifying the relationship between nodes u and v and let $\mathbf{S} \in \mathbb{R}^{|\mathcal{V}| \times |\mathcal{V}|}$ denote the *similarity matrix* summarizing all the pairwise node statistics.

Even though there is no "machine learning" involved in any of the statistical measures discussed in this section, they are still very useful and powerful baselines for relation prediction. Given a neighborhood overlap statistic $\mathbf{S}[u, v]$, a common strategy is to assume that the likelihood of an edge (u, v) is simply proportional to $\mathbf{S}[u, v]$:

$$P(\mathbf{A}[u, v] = 1) \propto \mathbf{S}[u, v]. \tag{2.8}$$

Thus, in order to approach the relation prediction task using a neighborhood overlap measure, one simply needs to set a threshold to determine when to predict the existence of an edge. Note that in the relation prediction setting we generally assume that we only know a subset of the true edges $\mathcal{E}_{\text{train}} \subset \mathcal{E}$. Our hope is that node-node similarity measures computed on the training edges will lead to accurate predictions about the existence of test (i.e., unseen) edges (Figure 2.3).

2.2.1   LOCAL OVERLAP MEASURES

Local overlap statistics are simply functions of the number of common neighbors two nodes share, i.e., $|\mathcal{N}(u) \cap \mathcal{N}(v)|$. For instance, the Sorensen index defines a matrix $\mathbf{S}_{\text{Sorenson}} \in \mathbb{R}^{|\mathcal{V}| \times |\mathcal{V}|}$ of node-node neighborhood overlaps with entries given by

$$\mathbf{S}_{\text{Sorenson}}[u, v] = \frac{2|\mathcal{N}(u) \cap \mathcal{N}(v)|}{d_u + d_v}, \tag{2.9}$$

which normalizes the count of common neighbors by the sum of the node degrees. Normalization of some kind is usually very important; otherwise, the overlap measure would be highly biased toward predicting edges for nodes with large degrees. Other similar approaches include the Salton index, which normalizes by the product of the degrees of u and v, i.e.,

$$\mathbf{S}_{\text{Salton}}[u, v] = \frac{2|\mathcal{N}(u) \cap \mathcal{N}(v)|}{\sqrt{d_u d_v}}, \tag{2.10}$$

as well as the Jaccard overlap:

$$\mathbf{S}_{\text{Jaccard}}(u, v) = \frac{|\mathcal{N}(u) \cap \mathcal{N}(v)|}{|\mathcal{N}(u) \cup \mathcal{N}(v)|}. \tag{2.11}$$

In general, these measures seek to quantify the overlap between node neighborhoods while minimizing any biases due to node degrees. There are many further variations of this approach in the literature [Lü and Zhou, 2011].

There are also measures that go beyond simply counting the number of common neighbors and that seek to consider the *importance* of common neighbors in some way. The Resource Allocation (RA) index counts the inverse degrees of the common neighbors,

$$\mathbf{S}_{\text{RA}}[v_1, v_2] = \sum_{u \in \mathcal{N}(v_1) \cap \mathcal{N}(v_2)} \frac{1}{d_u}, \tag{2.12}$$

while the Adamic–Adar (AA) index performs a similar computation using the inverse logarithm of the degrees:

$$\mathbf{S}_{\text{AA}}[v_1, v_2] = \sum_{u \in \mathcal{N}(v_1) \cap \mathcal{N}(v_2)} \frac{1}{\log(d_u)}. \tag{2.13}$$

Both these measures give more weight to common neighbors that have low degree, with intuition that a shared low-degree neighbor is more informative than a shared high-degree one.

2.2.2   GLOBAL OVERLAP MEASURES

Local overlap measures are extremely effective heuristics for link prediction and often achieve competitive performance even compared to advanced deep learning approaches [Perozzi et al., 2014]. However, the local approaches are limited in that they only consider local node neighborhoods. For example, two nodes could have no local overlap in their neighborhoods but still be members of the same community in the graph. *Global overlap* statistics attempt to take such relationships into account.

Katz Index
The Katz index is the most basic global overlap statistic. To compute the Katz index we simply count the number of paths *of all lengths* between a pair of nodes:

$$\mathbf{S}_{\text{Katz}}[u, v] = \sum_{i=1}^{\infty} \beta^i \mathbf{A}^i [u, v], \tag{2.14}$$

where $\beta \in \mathbb{R}^+$ is a user-defined parameter controlling how much weight is given to short vs. long paths. A small value of $\beta < 1$ would down-weight the importance of long paths.

Geometric series of matrices   The Katz index is one example of a geometric series of matrices, variants of which occur frequently in graph analysis and graph representation learning. The solution to a basic geometric series of matrices is given by the following theorem:

Theorem 2.1   *Let* $\mathbf{X}$ *be a real-valued square matrix and let* λ_1 *denote the largest eigenvalue of* $\mathbf{X}$. *Then*

$$(\mathbf{I} - \mathbf{X})^{-1} = \sum_{i=0}^{\infty} \mathbf{X}^i$$

if and only if $\lambda_1 < 1$ *and* $(\mathbf{I} - \mathbf{X})$ *is non-singular.*

Proof.   Let $s_n = \sum_{i=0}^{n} \mathbf{X}^i$ then we have that

$$\mathbf{X}s_n = \mathbf{X} \sum_{i=0}^{n} \mathbf{X}^i$$

$$= \sum_{i=1}^{n+1} \mathbf{X}^i$$

and

$$s_n - \mathbf{X}s_n = \sum_{i=0}^{n} \mathbf{X}^i - \sum_{i=1}^{n+1} \mathbf{X}^i$$
$$s_n(\mathbf{I} - \mathbf{X}) = \mathbf{I} - \mathbf{X}^{n+1}$$
$$s_n = (\mathbf{I} - \mathbf{X}^{n+1})(\mathbf{I} - \mathbf{X})^{-1}$$

And if $\lambda_1 < 1$ we have that $\lim_{n\to\infty} \mathbf{X}^n = 0$ so

$$\lim_{n\to\infty} s_n = \lim_{n\to\infty} (\mathbf{I} - \mathbf{X}^{n+1})(\mathbf{I} - \mathbf{X})^{-1}$$
$$= \mathbf{I}(\mathbf{I} - \mathbf{X})^{-1}$$
$$= (\mathbf{I} - \mathbf{X})^{-1}.$$

□

Based on Theorem 2.1, we can see that the solution to the Katz index is given by

$$\mathbf{S}_{\text{Katz}} = (\mathbf{I} - \beta\mathbf{A})^{-1} - \mathbf{I}, \tag{2.15}$$

where $\mathbf{S}_{\text{Katz}} \in \mathbb{R}^{|\mathcal{V}| \times |\mathcal{V}|}$ is the full matrix of node-node similarity values.

Leicht, Holme, and Newman (LHN) Similarity

One issue with the Katz index is that it is strongly biased by node degree. Equation (2.14) is generally going to give higher overall similarity scores when considering high-degree nodes, compared to low-degree ones, since high-degree nodes will generally be involved in more paths. To alleviate this, Leicht et al. [2006] propose an improved metric by considering the ratio between the actual number of observed paths and the *number of expected paths between two nodes*:

$$\frac{\mathbf{A}^i}{\mathbb{E}[\mathbf{A}^i]}, \tag{2.16}$$

i.e., the number of paths between two nodes is normalized based on our expectation of how many paths we expect under a random model.

To compute the expectation $\mathbb{E}[\mathbf{A}^i]$, we rely on what is called the *configuration model*, which assumes that we draw a random graph with the same set of degrees as our given graph. Under this assumption, we can analytically compute that

$$\mathbb{E}[\mathbf{A}[u, v]] = \frac{d_u d_v}{2m}, \tag{2.17}$$

where we have used $m = |\mathcal{E}|$ to denote the total number of edges in the graph. Equation (2.17) states that under a random configuration model, the likelihood of an edge is simply proportional to the product of the two node degrees. This can be seen by noting that there are d_u edges leaving u and each of these edges has a $\frac{d_v}{2m}$ chance of ending at v. For $\mathbb{E}[\mathbf{A}^2[u, v]]$ we can similarly compute

$$\mathbb{E}[\mathbf{A}^2[v_1, v_2]] = \frac{d_{v_1} d_{v_2}}{(2m)^2} \sum_{u \in \mathcal{V}} (d_u - 1) d_u. \tag{2.18}$$

This follows from the fact that path of length 2 could pass through any intermediate vertex u, and the likelihood of such a path is proportional to the likelihood that an edge leaving v_1 hits u—given by $\frac{d_{v_1} d_u}{2m}$—multiplied by the probability that an edge leaving u hits v_2—given by $\frac{d_{v_2}(d_u - 1)}{2m}$ (where we subtract one since we have already used up one of u's edges for the incoming edge from v_1).

Unfortunately the analytical computation of expected node path counts under a random configuration model becomes intractable as we go beyond paths of length three. Thus, to obtain the expectation $\mathbb{E}[\mathbf{A}^i]$ for longer path lengths (i.e., $i > 2$), Leicht et al. [2006] rely on the fact the largest eigenvalue can be used to approximate the growth in the number of paths. In particular, if we define $\mathbf{p}_i \in \mathbb{R}^{|\mathcal{V}|}$ as the vector counting the number of length-i paths between node u and all other nodes, then we have that for large i

$$\mathbf{A}\mathbf{p}_i = \lambda_1 \mathbf{p}_{i-1}, \tag{2.19}$$

since $\mathbf{p}_i$ will eventually converge to the dominant eigenvector of the graph. This implies that the number of paths between two nodes grows by a factor of λ_1 at each iteration, where we recall that λ_1 is the largest eigenvalue of $\mathbf{A}$. Based on this approximation for large i as well as the exact solution for $i = 1$ we obtain:

$$\mathbb{E}[\mathbf{A}^i[u, v]] = \frac{d_u d_v \lambda^{i-1}}{2m}. \tag{2.20}$$

Finally, putting it all together we can obtain a normalized version of the Katz index, which we term the LNH index (based on the initials of the authors who proposed the algorithm):

$$\mathbf{S}_{\mathrm{LNH}}[u, v] = \mathbf{I}[u, v] + \frac{2m}{d_u d_v} \sum_{i=0}^{\infty} \beta \lambda_1^{1-i} \mathbf{A}^i[u, v], \tag{2.21}$$

where $\mathbf{I}$ is a $|\mathcal{V}| \times |\mathcal{V}|$ identity matrix indexed in a consistent manner as $\mathbf{A}$. Unlike the Katz index, the LNH index accounts for the *expected* number of paths between nodes and only gives a high similarity measure if two nodes occur on more paths than we expect. Using Theorem 2.1 the solution to the matrix series (after ignoring diagonal terms) can be written as [Lü and Zhou, 2011]:

$$\mathbf{S}_{\mathrm{LNH}} = 2\alpha m \lambda_1 \mathbf{D}^{-1} \left(\mathbf{I} - \frac{\beta}{\lambda_1} \mathbf{A} \right)^{-1} \mathbf{D}^{-1}, \tag{2.22}$$

where $\mathbf{D}$ is a matrix with node degrees on the diagonal.

Random Walk Methods

Another set of global similarity measures consider random walks rather than exact counts of paths over the graph. For example, we can directly apply a variant of the famous PageRank approach [Page et al., 1999][4]—known as the Personalized PageRank algorithm [Leskovec et al., 2020]—where we define the stochastic matrix $\mathbf{P} = \mathbf{A}\mathbf{D}^{-1}$ and compute:

$$\mathbf{q}_u = c\mathbf{P}\mathbf{q}_u + (1-c)\mathbf{e}_u. \tag{2.23}$$

In this equation $\mathbf{e}_u$ is a one-hot indicator vector for node u and $\mathbf{q}_u[v]$ gives the stationary probability that random walk starting at node u visits node v. Here, the c term determines the probability that the random walk restarts at node u at each timestep. Without this restart probability, the random walk probabilities would simply converge to a normalized variant of the eigenvector centrality. However, with this restart probability we instead obtain a measure of importance specific to the node u, since the random walks are continually being "teleported" back to that node. The solution to this recurrence is given by

$$\mathbf{q}_u = (1-c)(\mathbf{I} - c\mathbf{P})^{-1}\mathbf{e}_u, \tag{2.24}$$

and we can define a node-node random walk similarity measure as

$$\mathbf{S}_{\text{RW}}[u, v] = \mathbf{q}_u[v] + \mathbf{q}_v[u], \tag{2.25}$$

i.e., the similarity between a pair of nodes is proportional to how likely we are to reach each node from a random walk starting from the other node.

2.3   GRAPH LAPLACIANS AND SPECTRAL METHODS

Having discussed traditional approaches to classification with graph data (Section 2.1) as well as traditional approaches to relation prediction (Section 2.2), we now turn to the problem of learning to cluster the nodes in a graph. This section will also motivate the task of learning low dimensional embeddings of nodes. We begin with the definition of some important matrices that can be used to represent graphs and a brief introduction to the foundations of *spectral graph theory*.

2.3.1   GRAPH LAPLACIANS

Adjacency matrices can represent graphs without any loss of information. However, there are alternative matrix representations of graphs that have useful mathematical properties. These matrix representations are called *Laplacians* and are formed by various transformations of the adjacency matrix.

---

[4]PageRank was developed by the founders of Google and powered early versions of the search engine.

Unnormalized Laplacian

The most basic Laplacian matrix is the unnormalized Laplacian, defined as follows:

$$\mathbf{L} = \mathbf{D} - \mathbf{A}, \tag{2.26}$$

where $\mathbf{A}$ is the adjacency matrix and $\mathbf{D}$ is the degree matrix. The Laplacian matrix of a simple graph has a number of important properties.

1. It is symmetric ($\mathbf{L}^T = \mathbf{L}$) and positive semi-definite ($\mathbf{x}^T \mathbf{L} \mathbf{x} \geq 0, \forall \mathbf{x} \in \mathbb{R}^{|\mathcal{V}|}$).

2. The following vector identity holds $\forall \mathbf{x} \in \mathbb{R}^{|\mathcal{V}|}$

$$\mathbf{x}^T \mathbf{L} \mathbf{x} = \frac{1}{2} \sum_{u \in \mathcal{V}} \sum_{v \in \mathcal{V}} \mathbf{A}[u, v](\mathbf{x}[u] - \mathbf{x}[v])^2 \tag{2.27}$$

$$= \sum_{(u,v) \in \mathcal{E}} (\mathbf{x}[u] - \mathbf{x}[v])^2. \tag{2.28}$$

3. $\mathbf{L}$ has $|V|$ non-negative eigenvalues: $0 = \lambda_{|\mathcal{V}|} \leq \lambda_{|\mathcal{V}|-1} \leq \dots \leq \lambda_1$.

The Laplacian and connected components The Laplacian summarizes many important properties of the graph. For example, we have the following theorem:

Theorem 2.2 *The geometric multiplicity of the 0 eigenvalue of the Laplacian $\mathbf{L}$ corresponds to the number of connected components in the graph.*

Proof. This can be seen by noting that for any eigenvector $\mathbf{e}$ of the eigenvalue 0 we have that

$$\mathbf{e}^T \mathbf{L} \mathbf{e} = 0 \tag{2.29}$$

by the definition of the eigenvalue-eigenvector equation. And, the result in Equation (2.29) implies that

$$\sum_{(u,v) \in \mathcal{E}} (\mathbf{e}[u] - \mathbf{e}[v])^2 = 0. \tag{2.30}$$

The equality above then implies that $\mathbf{e}[u] = \mathbf{e}[v], \forall (u, v) \in \mathcal{E}$, which in turn implies that $\mathbf{e}[u]$ is the same constant for all nodes u that are in the same connected component. Thus, if the graph is fully connected then the eigenvector for the eigenvalue 0 will be a constant vector of ones for all nodes in the graph, and this will be the only eigenvector for eigenvalue 0, since in this case there is only one unique solution to Equation (2.29).

Conversely, if the graph is composed of multiple connected components then we will have that Equation (2.29) holds independently on each of the blocks of the Laplacian corresponding to each connected component. That is, if the graph is composed of K connected components, then there exists an ordering of the nodes in the graph such that the Laplacian matrix can be written as

$$\mathbf{L} = \begin{bmatrix} \mathbf{L}_1 & & & \\ & \mathbf{L}_2 & & \\ & & \ddots & \\ & & & \mathbf{L}_K \end{bmatrix}, \tag{2.31}$$

where each of the $\mathbf{L}_k$ blocks in this matrix is a valid graph Laplacian of a fully connected subgraph of the original graph. Since they are valid Laplacians of fully connected graphs, for each of the $\mathbf{L}_k$ blocks we will have that Equation (2.29) holds and that each of these sub-Laplacians has an eigenvalue of 0 with multiplicity 1 and an eigenvector of all ones (defined only over the nodes in that component). Moreover, since $\mathbf{L}$ is a block diagonal matrix, its spectrum is given by the union of the spectra of all the $\mathbf{L}_k$ blocks, i.e., the eigenvalues of $\mathbf{L}$ are the union of the eigenvalues of the $\mathbf{L}_k$ matrices and the eigenvectors of $\mathbf{L}$ are the union of the eigenvectors of all the $\mathbf{L}_k$ matrices with 0 values filled at the positions of the other blocks. Thus, we can see that each block contributes one eigenvector for eigenvalue 0, and this *eigenvector is an indicator vector for the nodes in that connected component.*   □

Normalized Laplacians

In addition to the unnormalized Laplacian there are also two popular normalized variants of the Laplacian. The symmetric normalized Laplacian is defined as

$$\mathbf{L}_{\text{sym}} = \mathbf{D}^{-\frac{1}{2}} \mathbf{L} \mathbf{D}^{-\frac{1}{2}}, \tag{2.32}$$

while the random walk Laplacian is defined as

$$\mathbf{L}_{\text{RW}} = \mathbf{D}^{-1} \mathbf{L}. \tag{2.33}$$

Both of these matrices have similar properties as the Laplacian, but their algebraic properties differ by small constants due to the normalization. For example, Theorem 2.2 holds exactly for $\mathbf{L}_{\text{RW}}$. For $\mathbf{L}_{\text{sym}}$, Theorem 2.2 holds but with the eigenvectors for the 0 eigenvalue scaled by $\mathbf{D}^{\frac{1}{2}}$. As we will see throughout this book, these different variants of the Laplacian can be useful for different analysis and learning tasks.

2.3.2   GRAPH CUTS AND CLUSTERING

In Theorem 2.2, we saw that the eigenvectors corresponding to the 0 eigenvalue of the Laplacian can be used to assign nodes to clusters based on which connected component they belong to. However, this approach only allows us to cluster nodes that are already in disconnected components, which is trivial. In this section, we take this idea one step further and show that the Laplacian can be used to give an optimal clustering of nodes *within a fully connected graph*.

Graph Cuts

In order to motivate the Laplacian spectral clustering approach, we first must define what we mean by an *optimal* cluster. To do so, we define the notion of a *cut* on a graph. Let $\mathcal{A} \subset \mathcal{V}$ denote a subset of the nodes in the graph and let $\bar{\mathcal{A}}$ denote the complement of this set, i.e., $\mathcal{A} \cup \bar{\mathcal{A}} = \mathcal{V}, \mathcal{A} \cap \bar{\mathcal{A}} = \emptyset$. Given a partitioning of the graph into K non-overlapping subsets $\mathcal{A}_1, ..., \mathcal{A}_K$ we define the cut value of this partition as

$$\text{cut}(\mathcal{A}_1, ..., \mathcal{A}_K) = \frac{1}{2} \sum_{k=1}^{K} |(u, v) \in \mathcal{E} : u \in \mathcal{A}_k, v \in \bar{\mathcal{A}}_k|. \tag{2.34}$$

In other words, the cut is simply the count of how many edges cross the boundary between the partition of nodes. Now, one option to define an *optimal clustering* of the nodes into K clusters would be to select a partition that minimizes this cut value. There are efficient algorithms to solve this task, but a known problem with this approach is that it tends to simply make clusters that consist of a single node [Stoer and Wagner, 1997].

Thus, instead of simply minimizing the cut we generally seek to minimize the cut while also enforcing that the partitions are all reasonably large. One popular way of enforcing this is by minimizing the *Ratio Cut*:

$$\text{RatioCut}(\mathcal{A}_1, ..., \mathcal{A}_K) = \frac{1}{2} \frac{\sum_{k=1}^{K} |(u, v) \in \mathcal{E} : u \in \mathcal{A}_k, v \in \bar{\mathcal{A}}_k|}{|\mathcal{A}_k|}, \tag{2.35}$$

which penalizes the solution for choosing small cluster sizes. Another popular solution is to minimize the *Normalized Cut (NCut)*:

$$\text{NCut}(\mathcal{A}_1, ..., \mathcal{A}_K) = \frac{1}{2} \frac{\sum_{k=1}^{K} |(u, v) \in \mathcal{E} : u \in \mathcal{A}_k, v \in \bar{\mathcal{A}}_k|}{\text{vol}(\mathcal{A}_k)}, \tag{2.36}$$

where $\text{vol}(\mathcal{A}) = \sum_{u \in \mathcal{A}} d_u$. The NCut enforces that all clusters have a similar number of edges incident to their nodes.

Approximating the RatioCut with the Laplacian Spectrum

We will now derive an approach to find a cluster assignment that minimizes the RatioCut using the Laplacian spectrum. (A similar approach can be used to minimize the NCut value as well.) For simplicity, we will consider the case where we $K = 2$ and we are separating our nodes into two clusters. Our goal is to solve the following optimization problem:

$$\min_{\mathcal{A} \in \mathcal{V}} \text{RatioCut}(\mathcal{A}, \bar{\mathcal{A}}). \tag{2.37}$$

To rewrite this problem in a more convenient way, we define the following vector $\mathbf{a} \in \mathbb{R}^{|\mathcal{V}|}$:

$$\mathbf{a}[u] = \begin{cases} \sqrt{\frac{|\bar{\mathcal{A}}|}{|\mathcal{A}|}} & \text{if } u \in \mathcal{A} \\ -\sqrt{\frac{|\mathcal{A}|}{|\bar{\mathcal{A}}|}} & \text{if } u \in \bar{\mathcal{A}} \end{cases}. \tag{2.38}$$

Combining this vector with our properties of the graph Laplacian we can see that

$$\mathbf{a}^{\top}\mathbf{L}\mathbf{a} = \sum_{(u,v)\in\mathcal{E}} (\mathbf{a}[u] - \mathbf{a}[v])^2 \tag{2.39}$$

$$= \sum_{(u,v)\in\mathcal{E} \,:\, u\in\mathcal{A},v\in\bar{\mathcal{A}}} (\mathbf{a}[u] - \mathbf{a}[v])^2 \tag{2.40}$$

$$= \sum_{(u,v)\in\mathcal{E} \,:\, u\in\mathcal{A},v\in\bar{\mathcal{A}}} \left(\sqrt{\frac{|\bar{\mathcal{A}}|}{|\mathcal{A}|}} - \left(-\sqrt{\frac{|\mathcal{A}|}{|\bar{\mathcal{A}}|}} \right) \right)^2 \tag{2.41}$$

$$= \mathrm{cut}(\mathcal{A}, \bar{\mathcal{A}}) \left(\frac{|\mathcal{A}|}{|\bar{\mathcal{A}}|} + \frac{|\bar{\mathcal{A}}|}{|\mathcal{A}|} + 2 \right) \tag{2.42}$$

$$= \mathrm{cut}(\mathcal{A}, \bar{\mathcal{A}}) \left(\frac{|\mathcal{A}| + |\bar{\mathcal{A}}|}{|\bar{\mathcal{A}}|} + \frac{|\mathcal{A}| + |\bar{\mathcal{A}}|}{|\mathcal{A}|} \right) \tag{2.43}$$

$$= |\mathcal{V}|\mathrm{RatioCut}(\mathcal{A}, \bar{\mathcal{A}}). \tag{2.44}$$

Thus, we can see that $\mathbf{a}$ allows us to write the Ratio Cut in terms of the Laplacian (up to a constant factor). In addition, $\mathbf{a}$ has two other important properties:

$$\sum_{u\in\mathcal{V}} \mathbf{a}[u] = 0 \Leftrightarrow \mathbf{a} \perp \mathbf{1} \qquad \text{(Property 1)} \tag{2.45}$$

$$\|\mathbf{a}\|^2 = |\mathcal{V}| \qquad \text{(Property 2)}, \tag{2.46}$$

where $\mathbf{1}$ is the vector of all ones.

Putting this all together we can rewrite the Ratio Cut minimization problem in Equation (2.37) as

$$\min_{\mathcal{A}\in\mathcal{V}} \mathbf{a}^{\top}\mathbf{L}\mathbf{a} \tag{2.47}$$
$$s.t.$$
$$\mathbf{a} \perp \mathbf{1}$$
$$\|\mathbf{a}\|^2 = |\mathcal{V}|$$
$$\mathbf{a} \text{ defined as in Equation (2.38)}.$$

Unfortunately, however, this is an NP-hard problem since the restriction that $\mathbf{a}$ is defined as in Equation (2.38) requires that we are optimizing over a discrete set. The obvious relaxation is to remove this discreteness condition and simplify the minimization to be over real-valued vectors:

$$\min_{\mathbf{a} \in \mathbb{R}^{|\mathcal{V}|}} \mathbf{a}^{\top} \mathbf{La} \tag{2.48}$$

$$s.t.$$

$$\mathbf{a} \perp \mathbf{1}$$

$$\|\mathbf{a}\|^2 = |\mathcal{V}|.$$

By the Rayleigh–Ritz Theorem, the solution to this optimization problem is given by the second-smallest eigenvector of $\mathbf{L}$ (since the smallest eigenvector is equal to $\mathbf{1}$).

Thus, we can approximate the minimization of the RatioCut by setting $\mathbf{a}$ to be the second-smallest eigenvector[5] of the Laplacian. To turn this real-valued vector into a set of discrete cluster assignments, we can simply assign nodes to clusters based on the sign of $\mathbf{a}[u]$, i.e.,

$$\begin{cases} u \in \mathcal{A} & \text{if } \mathbf{a}[u] \geq 0 \\ u \in \bar{\mathcal{A}} & \text{if } \mathbf{a}[u] < 0. \end{cases} \tag{2.49}$$

In summary, the second-smallest eigenvector of the Laplacian is a continuous approximation to the discrete vector that gives an optimal cluster assignment (with respect to the RatioCut). An analogous result can be shown for approximating the NCut value, but it relies on the second-smallest eigenvector of the normalized Laplacian $\mathbf{L}_{\text{RW}}$ [Von Luxburg, 2007].

2.3.3 GENERALIZED SPECTRAL CLUSTERING

In the last section we saw that the spectrum of the Laplacian allowed us to find a meaningful partition of the graph into two clusters. In particular, we saw that the second-smallest eigenvector could be used to partition the nodes into different clusters. This general idea can be extended to an arbitrary number of K clusters by examining the K smallest eigenvectors of the Laplacian. The steps of this general approach are as follows:

1. Find the K smallest eigenvectors of $\mathbf{L}$ (excluding the smallest):

   $\mathbf{e}_{|\mathcal{V}|-1}, \mathbf{e}_{|\mathcal{V}|-2}, ..., \mathbf{e}_{|\mathcal{V}|-K}.$

2. Form the matrix $\mathbf{U} \in \mathbb{R}^{|\mathcal{V}| \times (K-1)}$ with the eigenvectors from Step 1 as columns.

3. Represent each node by its corresponding row in the matrix $\mathbf{U}$, i.e.,

   $$\mathbf{z}_u = \mathbf{U}[u] \ \forall u \in \mathcal{V}.$$

4. Run K-means clustering on the *embeddings* $\mathbf{z}_u \ \forall u \in \mathcal{V}$.

---

[5]Note that by second-smallest eigenvector we mean the eigenvector corresponding to the second-smallest eigenvalue.

As with the discussion of the $K = 2$ case in the previous section, this approach can be adapted to use the normalized Laplacian, and the approximation result for $K = 2$ can also be generalized to this $K > 2$ case [Von Luxburg, 2007].

The general principle of spectral clustering is a powerful one. We can represent the nodes in a graph using the spectrum of the graph Laplacian, and this representation can be motivated as a principled approximation to an optimal graph clustering. There are also close theoretical connections between spectral clustering and random walks on graphs, as well as the field of graph signal processing Ortega et al. [2018]. We will discuss many of these connections in future chapters.

2.4   TOWARD LEARNED REPRESENTATIONS

In the previous sections, we saw a number of traditional approaches to learning over graphs. We discussed how graph statistics and kernels can extract feature information for classification tasks. We saw how neighborhood overlap statistics can provide powerful heuristics for relation prediction. And, we offered a brief introduction to the notion of spectral clustering, which allows us to cluster nodes into communities in a principled manner. However, the approaches discussed in in this chapter—and especially the node and graph-level statistics—are limited due to the fact that they require careful, hand-engineered statistics and measures. These hand-engineered features are inflexible—i.e., they cannot adapt through a learning process—and designing these features can be a time-consuming and expensive process. The following chapters in this book introduce alternative approach to learning over graphs: *graph representation learning*. Instead of extracting hand-engineered features, we will seek to *learn* representations that encode structural information about the graph.

PART I

Node Embeddings

CHAPTER 3

Neighborhood Reconstruction Methods

This part of the book is concerned with methods for learning *node embeddings*. The goal of these methods is to encode nodes as low-dimensional vectors that summarize their graph position and the structure of their local graph neighborhood. In other words, we want to project nodes into a latent space, where geometric relations in this latent space correspond to relationships (e.g., edges) in the original graph or network [Hoff et al., 2002] (Figure 3.1).

In this chapter we will provide an overview of node embedding methods for simple and weighted graphs. Chapter 4 will provide an overview of analogous embedding approaches for multi-relational graphs.

3.1 AN ENCODER-DECODER PERSPECTIVE

We organize our discussion of node embeddings based upon the framework of *encoding and decoding* graphs. This way of viewing graph representation learning will reoccur throughout the book, and our presentation of node embedding methods based on this perspective closely follows Hamilton et al. [2017a].

In the encoder-decoder framework, we view the graph representation learning problem as involving two key operations. First, an *encoder* model maps each node in the graph into a low-dimensional vector or embedding. Next, a *decoder* model takes the low-dimensional node embeddings and uses them to reconstruct information about each node's neighborhood in the original graph. This idea is summarized in Figure 3.2.

3.1.1 THE ENCODER

Formally, the *encoder* is a function that maps nodes $v \in \mathcal{V}$ to vector embeddings $\mathbf{z}_v \in \mathbb{R}^d$ (where $\mathbf{z}_v$ corresponds to the embedding for node $v \in \mathcal{V}$). In the simplest case, the encoder has the following signature:

$$\text{ENC} : \mathcal{V} \to \mathbb{R}^d, \tag{3.1}$$

meaning that the encoder takes node IDs as input to generate the node embeddings. In most work on node embeddings, the encoder relies on what we call the *shallow embedding* approach, where this encoder function is simply an embedding lookup based on the node ID. In other

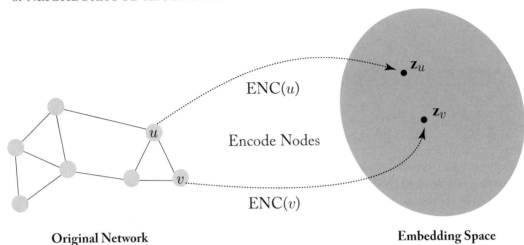

Figure 3.1: Illustration of the node embedding problem. Our goal is to learn an encoder (ENC), which maps nodes to a low-dimensional embedding space. These embeddings are optimized so that distances in the embedding space reflect the relative positions of the nodes in the original graph.

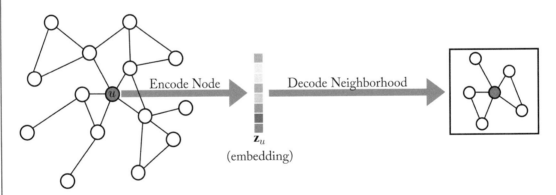

Figure 3.2: Overview of the encoder-decoder approach. The encoder maps the node u to a low-dimensional embedding $\mathbf{z}_u$. The decoder then uses $\mathbf{z}_u$ to reconstruct u's local neighborhood information.

words, we have that

$$\text{ENC}(v) = \mathbf{Z}[v], \tag{3.2}$$

where $\mathbf{Z} \in \mathbb{R}^{|\mathcal{V}| \times d}$ is a matrix containing the embedding vectors for all nodes and $\mathbf{Z}[v]$ denotes the row of $\mathbf{Z}$ corresponding to node v.

Shallow embedding methods will be the focus of this chapter. However, we note that the encoder can also be generalized beyond the shallow embedding approach. For instance, the encoder can use node features or the local graph structure around each node as input to generate an embedding. These generalized encoder architectures—often called graph neural networks (GNNs)—will be the main focus of Part II of this book.

3.1.2   THE DECODER

The role of the *decoder* is to reconstruct certain graph statistics from the node embeddings that are generated by the encoder. For example, given a node embedding $\mathbf{z}_u$ of a node u, the decoder might attempt to predict u's set of neighbors $\mathcal{N}(u)$ or its row $\mathbf{A}[u]$ in the graph adjacency matrix.

While many decoders are possible, the standard practice is to define *pairwise* decoders, which have the following signature:

$$\text{DEC} : \mathbb{R}^d \times \mathbb{R}^d \to \mathbb{R}^+. \tag{3.3}$$

Pairwise decoders can be interpreted as predicting the relationship or similarity between pairs of nodes. For instance, a simple pairwise decoder could predict whether two nodes are neighbors in the graph.

Applying the pairwise decoder to a pair of embeddings $(\mathbf{z}_u, \mathbf{z}_v)$ results in the *reconstruction* of the relationship between nodes u and v. The goal is optimize the encoder and decoder to minimize the reconstruction loss so that

$$\text{DEC}(\text{ENC}(u), \text{ENC}(v)) = \text{DEC}(\mathbf{z}_u, \mathbf{z}_v) \approx \mathbf{S}[u, v]. \tag{3.4}$$

Here, we assume that $\mathbf{S}[u, v]$ is a graph-based similarity measure between nodes. For example, the simple reconstruction objective of predicting whether two nodes are neighbors would correspond to $\mathbf{S}[u, v] \triangleq \mathbf{A}[u, v]$. However, one can define $\mathbf{S}[u, v]$ in more general ways as well, for example, by leveraging any of the pairwise neighborhood overlap statistics discussed in Section 2.2.

3.1.3   OPTIMIZING AN ENCODER-DECODER MODEL

To achieve the reconstruction objective (Equation (3.4)), the standard practice is to minimize an empirical reconstruction loss $\mathcal{L}$ over a set of training node pairs $\mathcal{D}$:

$$\mathcal{L} = \sum_{(u,v)\in\mathcal{D}} \ell \left(\text{DEC}(\mathbf{z}_u, \mathbf{z}_v), \mathbf{S}[u, v] \right), \tag{3.5}$$

where $\ell : \mathbb{R} \times \mathbb{R} \to \mathbb{R}$ is a loss function measuring the discrepancy between the decoded (i.e., estimated) similarity values $\text{DEC}(\mathbf{z}_u, \mathbf{z}_v)$ and the true values $\mathbf{S}[u, v]$. Depending on the definition of the decoder (DEC) and similarity function ($\mathbf{S}$), the loss function ℓ might be a mean-squared error or even a classification loss, such as cross entropy. Thus, the overall objective is to train the

Table 3.1: A summary of some well-known shallow embedding embedding algorithms. Note that the decoders and similarity functions for the random-walk based methods are asymmetric, with the similarity function $p_{\mathcal{G}}(v|u)$ corresponding to the probability of visiting v on a fixed-length random walk starting from u. Adapted from Hamilton et al. [2017a].

Method	Decoder	Similarity Measure	Loss Function
Lap Eigenmaps	$\|\mathbf{z}_u - \mathbf{z}_v\|_2^2$	General	$\mathrm{DEC}(\mathbf{z}_u, \mathbf{z}_v) \cdot \mathbf{S}[u, v]$
Graph Factorization	$\mathbf{z}_u^\top \mathbf{z}_v$	$\mathbf{A}[u, v]$	$\|\mathrm{DEC}(\mathbf{z}_u, \mathbf{z}_v) \cdot \mathbf{S}[u, v]\|_2^2$
GraRep	$\mathbf{z}_u^\top \mathbf{z}_v$	$\mathbf{A}[u, v],\ldots, \mathbf{A}^k[u, v]$	$\|\mathrm{DEC}(\mathbf{z}_u, \mathbf{z}_v) \cdot \mathbf{S}[u, v]\|_2^2$
HOPE	$\mathbf{z}_u^\top \mathbf{z}_v$	General	$\|\mathrm{DEC}(\mathbf{z}_u, \mathbf{z}_v) \cdot \mathbf{S}[u, v]\|_2^2$
DeepWalk	$\dfrac{e^{\mathbf{z}_u^\top \mathbf{z}_v}}{\sum_{k \in v} e^{\mathbf{z}_u^\top \mathbf{z}_k}}$	$p_{\mathcal{G}}(v|u)$	$-\mathbf{S}[u, v] \log(\mathrm{DEC}(\mathbf{z}_u, \mathbf{z}_v))$
node2vec	$\dfrac{e^{\mathbf{z}_u^\top \mathbf{z}_v}}{\sum_{k \in v} e^{\mathbf{z}_u^\top \mathbf{z}_k}}$	$p_{\mathcal{G}}(v|u)$ (biased)	$-\mathbf{S}[u, v] \log(\mathrm{DEC}(\mathbf{z}_u, \mathbf{z}_v))$

encoder and the decoder so that pairwise node relationships can be effectively reconstructed on the training set $\mathcal{D}$. Most approaches minimize the loss in Equation (3.5) using stochastic gradient descent [Robbins and Monro, 1951], but there are certain instances when more specialized optimization methods (e.g., based on matrix factorization) can be used.

3.1.4   OVERVIEW OF THE ENCODER-DECODER APPROACH

Table 3.1 applies this encoder-decoder perspective to summarize several well-known node embedding methods—all of which use the shallow encoding approach. The key benefit of the encoder-decoder framework is that it allows one to succinctly define and compare different embedding methods based on (i) their decoder function, (ii) their graph-based similarity measure, and (iii) their loss function.

In the following sections, we will describe the representative node embedding methods in Table 3.1 in more detail. We will begin with a discussion of node embedding methods that are motivated by matrix factorization approaches (Section 3.2) and that have close theoretical connections to spectral clustering (see Chapter 1). Following this, we will discuss more recent methods based on random walks (Section 3.3). These random walk approaches were initially motivated by inspirations from natural language processing, but—as we will discuss—they also share close theoretical ties to spectral graph theory.

3.2   FACTORIZATION-BASED APPROACHES

One way of viewing the encoder-decoder idea is from the perspective of matrix factorization. Indeed, the challenge of decoding local neighborhood structure from a node's embedding is closely related to reconstructing entries in the graph adjacency matrix. More generally, we can view this task as using matrix factorization to learn a low-dimensional approximation of a node-node similarity matrix $\mathbf{S}$, where $\mathbf{S}$ generalizes the adjacency matrix and captures some user-defined notion of node-node similarity (as discussed in Section 3.1.2) [Belkin and Niyogi, 2002, Kruskal, 1964].

Laplacian eigenmaps   One of the earliest—and most influential—factorization-based approaches is the Laplacian eigenmaps (LE) technique, which builds upon the spectral clustering ideas discussed in Chapter 2 [Belkin and Niyogi, 2002]. In this approach, we define the decoder based on the L2-distance between the node embeddings:

$$\text{DEC}(\mathbf{z}_u, \mathbf{z}_v) = \|\mathbf{z}_u - \mathbf{z}_v\|_2^2.$$

The loss function then weighs pairs of nodes according to their similarity in the graph:

$$\mathcal{L} = \sum_{(u,v) \in \mathcal{D}} \text{DEC}(\mathbf{z}_u, \mathbf{z}_v) \cdot \mathbf{S}[u, v]. \tag{3.6}$$

The intuition behind this approach is that Equation (3.6) penalizes the model when very similar nodes have embeddings that are far apart.

If $\mathbf{S}$ is constructed so that it satisfies the properties of a Laplacian matrix, then the node embeddings that minimize the loss in Equation (3.6) are identical to the solution for spectral clustering, which we discussed Section 2.3. In particular, if we assume the embeddings $\mathbf{z}_u$ are d-dimensional, then the optimal solution that minimizes Equation (3.6) is given by the d smallest eigenvectors of the Laplacian (excluding the eigenvector of all ones).

Inner-product methods   Following on the Laplacian eigenmaps technique, more recent work generally employs an inner-product based decoder, defined as follows:

$$\text{DEC}(\mathbf{z}_u, \mathbf{z}_v) = \mathbf{z}_u^\top \mathbf{z}_v. \tag{3.7}$$

Here, we assume that the similarity between two nodes—e.g., the overlap between their local neighborhoods—is proportional to the dot product of their embeddings.

Some examples of this style of node embedding algorithms include the Graph Factorization (GF) approach[1] [Ahmed et al., 2013], GraRep [Cao et al., 2015], and HOPE [Ou et al., 2016]. All three of these methods combine the inner-product decoder (Equation (3.7)) with the

---

[1]Of course, Ahmed et al. [2013] were not the first researchers to propose factorizing an adjacency matrix, but they were the first to present a scalable $O(|\mathcal{E}|)$ algorithm for the purpose of generating node embeddings.

following mean-squared error:

$$\mathcal{L} = \sum_{(u,v)\in\mathcal{D}} \|\text{DEC}(\mathbf{z}_u, \mathbf{z}_v) - \mathbf{S}[u, v]\|_2^2. \tag{3.8}$$

They differ primarily in how they define $\mathbf{S}[u, v]$, i.e., the notion of node-node similarity or neighborhood overlap that they use. Whereas the GF approach directly uses the adjacency matrix and sets $\mathbf{S} \triangleq \mathbf{A}$, the GraRep and HOPE approaches employ more general strategies. In particular, GraRep defines $\mathbf{S}$ based on powers of the adjacency matrix, while the HOPE algorithm supports general neighborhood overlap measures (e.g., any neighborhood overlap measure from Section 2.2).

These methods are referred to as matrix-factorization approaches, since their loss functions can be minimized using factorization algorithms, such as the singular value decomposition (SVD). Indeed, by stacking the node embeddings $\mathbf{z}_u \in \mathbb{R}^d$ into a matrix $\mathbf{Z} \in \mathbb{R}^{|\mathcal{V}|\times d}$ the reconstruction objective for these approaches can be written as

$$\mathcal{L} \approx \|\mathbf{Z}\mathbf{Z}^\top - \mathbf{S}\|_2^2, \tag{3.9}$$

which corresponds to a low-dimensional factorization of the node-node similarity matrix $\mathbf{S}$. Intuitively, the goal of these methods is to learn embeddings for each node such that the inner product between the learned embedding vectors approximates some deterministic measure of node similarity.

3.3   RANDOM WALK EMBEDDINGS

The inner-product methods discussed in the previous section all employ *deterministic* measures of node similarity. They often define $\mathbf{S}$ as some polynomial function of the adjacency matrix, and the node embeddings are optimized so that $\mathbf{z}_u^\top \mathbf{z}_v \approx \mathbf{S}[u, v]$. Building on these successes, recent years have seen a surge in successful methods that adapt the inner-product approach to use *stochastic* measures of neighborhood overlap. The key innovation in these approaches is that node embeddings are optimized so that two nodes have similar embeddings if they tend to co-occur on short random walks over the graph.

DeepWalk and node2vec   Similar to the matrix factorization approaches described above, DeepWalk and node2vec use a shallow embedding approach and an inner-product decoder. The key distinction in these methods is in how they define the notions of node similarity and neighborhood reconstruction. Instead of directly reconstructing the adjacency matrix $\mathbf{A}$—or some deterministic function of $\mathbf{A}$—these approaches optimize embeddings to encode the statistics of random walks. Mathematically, the goal is to learn embeddings so that the following (roughly) holds:

$$\mathrm{DEC}(\mathbf{z}_u, \mathbf{z}_v) \triangleq \frac{e^{\mathbf{z}_u^\top \mathbf{z}_v}}{\sum_{v_k \in \mathcal{V}} e^{\mathbf{z}_u^\top \mathbf{z}_k}} \tag{3.10}$$
$$\approx p_{\mathcal{G},T}(v|u),$$

where $p_{\mathcal{G},T}(v|u)$ is the probability of visiting v on a length-T random walk starting at u, with T usually defined to be in the range $T \in \{2, ..., 10\}$. Again, a key difference between Equation (3.10) and the factorization-based approaches (e.g., Equation (3.8)) is that the similarity measure in Equation (3.10) is both stochastic and asymmetric.

To train random walk embeddings, the general strategy is to use the decoder from Equation (3.10) and minimize the following cross-entropy loss:

$$\mathcal{L} = \sum_{(u,v) \in \mathcal{D}} -\log(\mathrm{DEC}(\mathbf{z}_u, \mathbf{z}_v)). \tag{3.11}$$

Here, we use $\mathcal{D}$ to denote the training set of random walks, which is generated by sampling random walks starting from each node. For example, we can assume that N pairs of co-occurring nodes for each node u are sampled from the distribution $(u, v) \sim p_{\mathcal{G},T}(v|u)$.

Unfortunately, however, naively evaluating the loss in Equation (3.11) can be computationally expensive. Indeed, evaluating the denominator in Equation (3.10) alone has time complexity $O(|\mathcal{V}|)$, which makes the overall time complexity of evaluating the loss function $O(|\mathcal{D}||\mathcal{V}|)$. There are different strategies to overcome this computational challenge, and this is one of the essential differences between the original DeepWalk and node2vec algorithms. DeepWalk employs a *hierarchical softmax* to approximate Equation (3.10), which involves leveraging a binary-tree structure to accelerate the computation [Perozzi et al., 2014]. On the other hand, node2vec employs a *noise contrastive* approach to approximate Equation (3.11), where the normalizing factor is approximated using *negative samples* in the following way [Grover and Leskovec, 2016]:

$$\mathcal{L} = \sum_{(u,v) \in \mathcal{D}} -\log(\sigma(\mathbf{z}_u^\top \mathbf{z}_v)) - \gamma \mathbb{E}_{v_n \sim P_n(\mathcal{V})}[\log(-\sigma(\mathbf{z}_u^\top \mathbf{z}_{v_n}))]. \tag{3.12}$$

Here, we use σ to denote the logistic function, $P_n(\mathcal{V})$ to denote a distribution over the set of nodes $\mathcal{V}$, and we assume that $\gamma > 0$ is a hyperparameter. In practice, $P_n(\mathcal{V})$ is often defined to be a uniform distribution, and the expectation is approximated using Monte Carlo sampling.

The node2vec approach also distinguishes itself from the earlier DeepWalk algorithm by allowing for a more flexible definition of random walks. In particular, whereas DeepWalk simply employs uniformly random walks to define $p_{\mathcal{G},T}(v|u)$, the node2vec approach introduces hyperparameters that allow the random walk probabilities to smoothly interpolate between walks that are more akin to breadth-first search or depth-first search over the graph.

Large-scale information network embeddings (LINE) In addition to DeepWalk and node2vec, Tang et al. [2015]'s LINE algorithm is often discussed within the context of random-walk approaches. The LINE approach does not explicitly leverage random walks, but it shares conceptual motivations with DeepWalk and node2vec. The basic idea in LINE is to combine two encoder-decoder objectives. The first objective aims to encode first-order adjacency information and uses the following decoder:

$$\text{DEC}(\mathbf{z}_u, \mathbf{z}_v) = \frac{1}{1 + e^{-\mathbf{z}_u^\top \mathbf{z}_v}}, \tag{3.13}$$

with an adjacency-based similarity measure (i.e., $\mathbf{S}[u, v] = \mathbf{A}[u, v]$). The second objective is more similar to the random walk approaches. It is the same decoder as Equation (3.10), but it is trained using the KL-divergence to encode two-hop adjacency information (i.e., the information in $\mathbf{A}^2$). Thus, LINE is conceptually related to node2vec and DeepWalk. It uses a probabilistic decoder and probabilistic loss function (based on the KL-divergence). However, instead of sampling random walks, it explicitly reconstructs first- and second-order neighborhood information.

Additional variants of the random-walk idea One benefit of the random walk approach is that it can be extended and modified by biasing or modifying the random walks. For example, Perozzi et al. [2016] consider random walks that "skip" over nodes, which generates a similarity measure similar to GraRep (discussed in Section 3.2), and Ribeiro et al. [2017] define random walks based on the structural relationships between nodes—rather than neighborhood information—which generates node embeddings that encode structural roles in the graph.

3.3.1 RANDOM WALK METHODS AND MATRIX FACTORIZATION

It can be shown that random walk methods are actually closely related to matrix factorization approaches [Qiu et al., 2018]. Suppose we define the following matrix of node-node similarity values:

$$\mathbf{S}_{\text{DW}} = \log\left(\frac{\text{vol}(\mathcal{V})}{T}\left(\sum_{t=1}^{T} \mathbf{P}^t\right) \mathbf{D}^{-1}\right) - \log(b), \tag{3.14}$$

where b is a constant and $\mathbf{P} = \mathbf{D}^{-1}\mathbf{A}$. In this case, Qiu et al. [2018] show that the embeddings $\mathbf{Z}$ learned by DeepWalk satisfy:

$$\mathbf{Z}^\top \mathbf{Z} \approx \mathbf{S}_{\text{DW}}. \tag{3.15}$$

Interestingly, we can also decompose the interior part of Equation (3.14) as

$$\left(\sum_{t=1}^{T} \mathbf{P}^t\right) \mathbf{D}^{-1} = \mathbf{D}^{-\frac{1}{2}}\left(\mathbf{U}\left(\sum_{t=1}^{T} \Lambda^t\right) \mathbf{U}^\top\right) \mathbf{D}^{-\frac{1}{2}}, \tag{3.16}$$

where $\mathbf{U}\Lambda\mathbf{U}^\top = \mathbf{L}_{\text{sym}}$ is the eigendecomposition of the symmetric normalized Laplacian. This reveals that the embeddings learned by DeepWalk are in fact closely related to the spectral clustering embeddings discussed in Part I of this book. The key difference is that the DeepWalk embeddings control the influence of different eigenvalues through T, i.e., the length of the random walk. Qiu et al. [2018] derive similar connections to matrix factorization for node2vec and discuss other related factorization-based approaches inspired by this connection.

3.4   LIMITATIONS OF SHALLOW EMBEDDINGS

This focus of this chapter—and this part of book more generally—has been on shallow embedding methods. In these approaches, the encoder model that maps nodes to embeddings is simply an embedding lookup (Equation (3.2)), which trains a unique embedding for each node in the graph. This approach has achieved many successes in the past decade, and in the next chapter we will discuss how this shallow approach can be generalized to multi-relational graphs. However, it is also important to note that shallow embedding approaches suffer from some important drawbacks.

1. The first issue is that shallow embedding methods do not share any parameters between nodes in the encoder, since the encoder directly optimizes a unique embedding vector for each node. This lack of parameter sharing is both statistically and computationally inefficient. From a statistical perspective, parameter sharing can improve the efficiency of learning and also act as a powerful form of regularization. From the computational perspective, the lack of parameter sharing means that the number of parameters in shallow embedding methods necessarily grows as $O(|\mathcal{V}|)$, which can be intractable in massive graphs.

2. A second key issue with shallow embedding approaches is that they do not leverage node features in the encoder. Many graph datasets have rich feature information, which could potentially be informative in the encoding process.

3. Last—and perhaps most importantly—shallow embedding methods are inherently *transductive* [Hamilton et al., 2017b]. These methods can only generate embeddings for nodes that were present during the training phase. Generating embeddings for new nodes—which are observed after the training phase—is not possible unless additional optimizations are performed to learn the embeddings for these nodes. This restriction prevents shallow embedding methods from being used on *inductive* applications, which involve generalizing to unseen nodes after training.

To alleviate these limitations, shallow encoders can be replaced with more sophisticated encoders that depend more generally on the structure and attributes of the graph. We will discuss the most popular paradigm to define such encoders—i.e., GNNs—in Part II of this book.

CHAPTER 4

Multi-Relational Data and Knowledge Graphs

In Chapter 3 we discussed approaches for learning low-dimensional embeddings of nodes. We focused on so-called *shallow embedding* approaches, where we learn a unique embedding for each node. In this chapter, we will continue our focus on shallow embedding methods, and we will introduce techniques to deal with multi-relational graphs.

Knowledge graph completion   Most of the methods we review in this chapter were originally designed for the task of knowledge graph completion. In knowledge graph completion, we are given a multi-relational graph $\mathcal{G} = (\mathcal{V}, \mathcal{E})$, where the edges are defined as tuples $e = (u, \tau, v)$ indicating the presence of a particular relation $\tau \in \mathcal{T}$ holding between two nodes. Such multi-relational graphs are often referred to as *knowledge graphs*, since we can interpret the tuple (u, τ, v) as specifying that a particular "fact" holds between the two nodes u and v. For example, in a biomedical knowledge graph we might have an edge type $\tau = \texttt{TREATS}$ and the edge $(u, \texttt{TREATS}, v)$ could indicate that the drug associated with node u treats the disease associated with node v. Generally, the goal in knowledge graph completion is to predict missing edges in the graph, i.e., relation prediction, but there are also examples of node classification tasks using multi-relational graphs [Schlichtkrull et al., 2017].

In this chapter we will provide a brief overview of embedding methods for multi-relational graphs, but it is important to note that a full treatment of knowledge graph completion is beyond the scope of this chapter. Not all knowledge graph completion methods rely on embeddings, and we will not cover every variation of embedding methods here. We refer interested readers to Nickel et al. [2016] for a complementary review of the area.

4.1   RECONSTRUCTING MULTI-RELATIONAL DATA

As with the simple graphs discussed in Chapter 3, we can view embedding multi-relational graphs as a reconstruction task. Given the embeddings $\mathbf{z}_u$ and $\mathbf{z}_v$ of two nodes, our goal is to reconstruct the relationship between these nodes. The complication—compared to the setting of the previous chapter—is that we now have to deal with the presence of multiple different types of edges.

To address this complication, we augment our decoder to make it multi-relational. Instead of only taking a pair of node embeddings as input, we now define the decoder as accepting a pair of node embeddings *as well as a relation type*, i.e., DEC : $\mathbb{R}^d \times \mathcal{R} \times \mathbb{R}^d \to \mathbb{R}^+$. We can interpret the output of this decoder, i.e., DEC($\mathbf{z}_u, \tau, \mathbf{z}_v$), as the likelihood that the edge (u, τ, v) exists in the graph.

To give a concrete example, one of the simplest and earliest approaches to learning multi-relational embeddings—often termed RESCAL—defined the decoder as [Nickel et al., 2011]:

$$\text{DEC}(u, \tau, v) = \mathbf{z}_u^\top \mathbf{R}_\tau \mathbf{z}_v, \tag{4.1}$$

where $\mathbf{R}_\tau \in \mathbb{R}^{d \times d}$ is a learnable matrix specific to relation $\tau \in \mathcal{R}$. Keeping things simple with this decoder, we could train our embedding matrix $\mathbf{Z}$ and our relation matrices $\mathbf{R}_\tau, \forall \tau \in \mathcal{R}$ using a basic reconstruction loss:

$$\mathcal{L} = \sum_{u \in \mathcal{V}} \sum_{v \in \mathcal{V}} \sum_{\tau \in \mathcal{R}} \|\text{DEC}(u, \tau, v) - \mathcal{A}[u, \tau, v]\|^2 \tag{4.2}$$

$$= \sum_{u \in \mathcal{V}} \sum_{v \in \mathcal{V}} \sum_{\tau \in \mathcal{R}} \|\mathbf{z}_u^\top \mathbf{R}_\tau \mathbf{z}_v - \mathcal{A}[u, \tau, v]\|^2, \tag{4.3}$$

where $\mathcal{A} \in \mathbb{R}^{|\mathcal{V}| \times |\mathcal{R}| \times |\mathcal{V}|}$ is the adjacency tensor for the multi-relational graph. If we were to optimize Equation (4.2), we would in fact be performing a kind of *tensor factorization*. This idea of factorizing a tensor thus generalizes the matrix factorization approaches discussed in Chapter 3.

Loss functions, decoders, and similarity functions   In Chapter 3 we discussed how the diversity of methods for node embeddings largely stem from the use of different decoders (DEC), similarity measures ($\mathbf{S}[u, v]$), and loss functions ($\mathcal{L}$). The decoder gives a score between a pair of node embeddings; the similarity function defines what kind of node-node similarity we are trying to decode; and the loss function tells us how to evaluate the discrepancy between the output of the decoder and the ground truth similarity measure.

In the multi-relational setting, we will also see a diversity of decoders and loss functions. However, nearly all multi-relational embedding methods simply define the similarity measure directly based on the adjacency tensor. In other words, all the methods in this chapter assume that we are trying to reconstruct immediate (multi-relational) neighbors from the low-dimensional embeddings. This is due to the difficulty of defining higher-order neighborhood relationships in multi-relational graphs, as well as the fact that most multi-relational embedding methods were specifically designed for relation prediction.

4.2   LOSS FUNCTIONS

As discussed above, the two key ingredients for a multi-relational node embedding method are the decoder and the loss function. We begin with a brief discussion of the standard loss functions used for this task, before turning our attention to the multitude of decoders that have been proposed in the literature.

As a motivation for the loss functions we consider, it is worth considering the drawbacks of the simple reconstruction loss we introduced in Equation (4.2). There are two major problems with this loss. The first issue is that it is extremely expensive to compute. The nested sums in Equation (4.2) require $O(|\mathcal{V}|^2||\mathcal{R}|)$ operations, and this computation time will be prohibitive in many large graphs. Moreover, since many multi-relational graphs are *sparse*—i.e., $|\mathcal{E}| \ll |\mathcal{V}|^2||\mathcal{R}|$—we would ideally want a loss function that is $O(|\mathcal{E}|)$. The second problem is more subtle. Our goal is to decode the adjacency tensor from the low-dimensional node embeddings. We know that (in most cases) this tensor will contain only binary values, but the mean-squared error in Equation (4.2) is not well suited to such a binary comparison. In fact, the mean-squared error is a natural loss for *regression* whereas our target is something closer to *classification on edges*.

Cross-Entropy with Negative Sampling

One popular loss function that is both efficient and suited to our task is the *cross-entropy loss with negative sampling*. We define this loss as:

$$\mathcal{L} = \sum_{(u,\tau,v)\in\mathcal{E}} -\log(\sigma(\text{DEC}(\mathbf{z}_u, \tau, \mathbf{z}_v))) - \gamma \mathbb{E}_{v_n \sim P_{n,u}(\mathcal{V})} \left[\log\left(\sigma\left(-\text{DEC}(\mathbf{z}_u, \tau, \mathbf{z}_{v_n})\right)\right)\right], \quad (4.4)$$

where σ denotes the logistic function, $P_{n,u}(\mathcal{V})$ denotes a "negative sampling" distribution over the set of nodes $\mathcal{V}$ (which might depend on u), and $\gamma > 0$ is a hyperparameter. This is essentially the same loss as we saw for node2vec (Equation (3.12)), but here we are considering general multi-relational decoders.

We call this a *cross-entropy* loss because it is derived from the standard binary cross-entropy loss. Since we are feeding the output of the decoder to a logistic function, we obtain normalized scores in [0, 1] that can be interpreted as probabilities. The term

$$\log(\sigma(\text{DEC}(\mathbf{z}_u, \tau, \mathbf{z}_v))) \quad (4.5)$$

then equals the log-likelihood that we predict "true" for an edge that does actually exist in the graph. On the other hand, the term

$$\mathbb{E}_{v_n \sim P_{n,u}(\mathcal{V})} \left[\log\left(\sigma\left(-\text{DEC}(\mathbf{z}_u, \tau, \mathbf{z}_{v_n})\right)\right)\right] \quad (4.6)$$

then equals the expected log-likelihood that we correctly predict "false" for an edge that does not exist in the graph. In practice, the expectation is evaluated using a Monte Carlo approximation and the most popular form of this loss is

$$\mathcal{L} = \sum_{(u,\tau,v)\in\mathcal{E}} -\log(\sigma(\text{DEC}(\mathbf{z}_u, \tau, \mathbf{z}_v))) - \sum_{v_n \in \mathcal{P}_{n,u}} \left[\log\left(\sigma\left(-\text{DEC}(\mathbf{z}_u, \tau, \mathbf{z}_{v_n})\right)\right)\right], \quad (4.7)$$

where $\mathcal{P}_{n,u}$ is a (usually small) set of nodes sampled from $P_{n,u}(\mathcal{V})$.

A note on negative sampling   The way in which negative samples are generated can have a large impact on the quality of the learned embeddings. The most common approach to define the distribution $P_{n,u}$ is to simply use a uniform distribution over all nodes in the graph. This is a simple strategy, but it also means that we will get "false negatives" in the cross-entropy calculation. In other words, it is possible that we accidentally sample a "negative" tuple (u, τ, v_n) that actually exists in the graph. Some works address this by filtering such false negatives.

Other variations of negative sampling attempt to produce more "difficult" negative samples. For example, some relations can only exist between certain types of nodes. (A node representing a person in a knowledge graph would be unlikely to be involved in an edge involving the MANUFACTURED-BY relation.) Thus, one strategy is to only sample negative examples that satisfy such type constraints. Sun et al. [2019] even propose an approach to select challenging negative samples by learning an adversarial model.

Note also that—without loss of generality—we have assumed that the negative sampling occurs over the second node in the edge tuple. That is, we assume that we draw a negative sample by replacing the *tail* node v in the tuple (u, τ, v) with a negative sample v_n. Always sampling the tail node simplifies notation but can lead to biases in multi-relational graphs where edge direction is important. In practice it can be better to draw negative samples for both the *head* node (i.e., u) and the tail node (i.e., v) of the relation.

Max-Margin Loss
The other popular loss function used for multi-relational node embedding is the margin loss:

$$\mathcal{L} = \sum_{(u,\tau,v)\in\mathcal{E}} \sum_{v_n\in\mathcal{P}_{n,u}} \max(0, -\text{DEC}(\mathbf{z}_u, \tau, \mathbf{z}_v) + \text{DEC}(\mathbf{z}_u, \tau, \mathbf{z}_{v_n}) + \Delta). \tag{4.8}$$

In this loss we are again comparing the decoded score for a true pair compared to a negative sample—a strategy often termed *contrastive estimation*. However, rather than treating this as a binary classification task, in Equation (4.8) we are simply comparing the direct output of the decoders. If the score for the "true" pair is bigger than the "negative" pair then we have a small loss. The Δ term is called the margin, and the loss will equal 0 if the difference in scores is at least that large for all examples. This loss is also known as the *hinge loss*.

4.3   MULTI-RELATIONAL DECODERS

The previous section introduced the two most popular loss functions for learning multi-learning node embeddings. These losses can be combined with various different decoder functions, and we turn our attention to the definition of these decoders now. So far, we have only discussed one possible multi-relational decoder, the so-called RESCAL decoder, which was introduced in

Table 4.1: Summary of some popular decoders used for multi-relational data

Name	Decoder	Relation Parameters
RESCAL	$\mathbf{z}_u^\top \mathbf{R}_\tau \mathbf{z}_v$	$\mathbf{R}_\tau \in \mathbb{R}^{d \times d}$
TransE	$-\|\mathbf{z}_u + \mathbf{r}_\tau - \mathbf{z}_v\|$	$\mathbf{r}_\tau \in \mathbb{R}^d$
TransX	$-\|g_{1,\tau}(\mathbf{z}_u) + \mathbf{r}_\tau - g_{2,\tau}(\mathbf{z}_v)\|$	$\mathbf{r}_\tau \in \mathbb{R}^d, g_{1,\tau}, g_{2,\tau} \in \mathbb{R}^d \rightarrow \mathbb{R}^d$
DistMult	$<\mathbf{z}_u, \mathbf{r}_\tau, \mathbf{z}_v>$	$\mathbf{r}_\tau \in \mathbb{R}^d$
ComplEx	$\mathrm{Re}\,(<\mathbf{z}_u, \mathbf{r}_\tau, \bar{\mathbf{z}}_v>)$	$\mathbf{r}_\tau \in \mathbb{C}^d$
RotatE	$-\|\mathbf{z}_u \circ \mathbf{r}_\tau - \mathbf{z}_v\|$	$\mathbf{r}_\tau \in \mathbb{C}^d$

Section 4.1:

$$\mathrm{DEC}(\mathbf{z}_u, \tau, \mathbf{z}_v) = \mathbf{z}_u^\top \mathbf{R}_\tau \mathbf{z}_v. \qquad (4.9)$$

In the RESCAL decoder, we associate a trainable matrix $\mathbf{R}_\tau \in \mathbb{R}^{d \times d}$ with each relation. However, one limitation of this approach—and a reason why it is not often used—is its high computational and statistical cost for representing relations. There are $O(d^2)$ parameters for each relation type in RESCAL, which means that relations require an order of magnitude more parameters to represent, compared to entities.

   More popular modern decoders aim to use only $O(d)$ parameters to represent each relation. We will discuss several popular variations of multi-relational decoders here, though our survey is far from exhaustive. The decoders surveyed in this chapter are summarized in Table 4.1.

Translational Decoders

One popular class of decoders represents relations as *translations* in the embedding space. This approach was initiated by Bordes et al. [2013]'s *TransE* model, which defined the decoder as

$$\mathrm{DEC}(\mathbf{z}_u, \tau, \mathbf{z}_v) = -\|\mathbf{z}_u + \mathbf{r}_\tau - \mathbf{z}_v\|. \qquad (4.10)$$

In these approaches, we represent each relation using a d-dimensional embedding. The likelihood of an edge is proportional to the distance between the embedding of the head node and the tail node, after *translating* the head node according to the relation embedding. TransE is one of the earliest multi-relational decoders proposed and continues to be a strong baseline in many applications.

   One limitation of TransE is its simplicity, however, and many works have also proposed extensions of this translation idea. We collectively refer to these models as *TransX* models and they have the form:

$$\mathrm{DEC}(\mathbf{z}_u, \tau, \mathbf{z}_v) = -\|g_{1,\tau}(\mathbf{z}_u) + \mathbf{r}_\tau - g_{2,\tau}(\mathbf{z}_v)\|, \qquad (4.11)$$

where $g_{i,\tau}$ are trainable transformations that depend on the relation τ. For example, Wang et al. [2014]'s *TransH* model defines the decoder as

$$\text{DEC}(\mathbf{z}_u, \tau, \mathbf{z}_v) = -\|(\mathbf{z}_u - \mathbf{w}_r^\top \mathbf{z}_u \mathbf{w}_r) + \mathbf{r}_\tau - (\mathbf{z}_u - \mathbf{w}_r^\top \mathbf{z}_v \mathbf{w}_r)\|. \tag{4.12}$$

The TransH approach projects the entity embeddings onto a learnable relation-specific hyperplane—defined by the normal vector $\mathbf{w}_r$—before performing translation. Additional variations of the TransE model are proposed in Nguyen et al. [2016] and Ji et al. [2015].

Multi-Linear Dot Products
Rather than defining a decoder based upon translating embeddings, a second popular line of work develops multi-relational decoders by generalizing the dot-product decoder from simple graphs. In this approach—often termed DistMult and first proposed by Yang et al.—we define the decoder as

$$\text{DEC}(\mathbf{z}_u, \tau, \mathbf{z}_v) = <\mathbf{z}_u, \mathbf{r}_\tau, \mathbf{z}_v> \tag{4.13}$$

$$= \sum_{i=1}^{d} \mathbf{z}_u[i] \times \mathbf{r}_\tau[i] \times \mathbf{z}_v[i]. \tag{4.14}$$

Thus, this approach takes a straightforward generalization of the dot product to be defined over three vectors.

Complex Decoders
One limitation of the DistMult decoder in Equation (4.13) is that it can only encode *symmetric* relations. In other words, for the multi-linear dot-product decoder defined in Equation (4.13), we have that

$$\text{DEC}(\mathbf{z}_u, \tau, \mathbf{z}_v) = <\mathbf{z}_u, \mathbf{r}_\tau, \mathbf{z}_v>$$
$$= \sum_{i=1}^{d} \mathbf{z}_u[i] \times \mathbf{r}_\tau[i] \times \mathbf{z}_v[i]$$
$$= <\mathbf{z}_v, \mathbf{r}_\tau, \mathbf{z}_u>$$
$$= \text{DEC}(\mathbf{z}_v, \tau, \mathbf{z}_u).$$

This is a serious limitation as many relation types in multi-relational graphs are directed and asymmetric. To address this issue, Trouillon et al. [2016] proposed augmenting the DistMult encoder by employing complex-valued embeddings. They define the *ComplEx* as

$$\text{DEC}(\mathbf{z}_u, \tau, \mathbf{z}_v) = \text{Re}(<\mathbf{z}_u, \mathbf{r}_\tau, \bar{\mathbf{z}}_v>) \tag{4.15}$$

$$= \text{Re}\left(\sum_{i=1}^{d} \mathbf{z}_u[i] \times \mathbf{r}_\tau[i] \times \bar{\mathbf{z}}_v[j]\right), \tag{4.16}$$

where now $\mathbf{z}_u, \mathbf{z}_v, \mathbf{r}_\tau \in \mathbb{C}^d$ are complex-valued embeddings and Re denotes the real component of a complex vector. Since we take the complex conjugate $\bar{\mathbf{z}}_v$ of the tail embedding, this approach to decoding can accommodate asymmetric relations.

A related approach, termed *RotatE*, defines the decoder as rotations in the complex plane as follows [Sun et al., 2019]:

$$\text{DEC}(\mathbf{z}_u, \tau, \mathbf{z}_v) = -\|\mathbf{z}_u \circ \mathbf{r}_\tau - \mathbf{z}_v\|, \tag{4.17}$$

where $\circ$ denotes the Hadamard product. In Equation (4.17) we again assume that all embeddings are complex valued, and we additionally constrain the entries of $\mathbf{r}_\tau$ so that $|\mathbf{r}_\tau[i]| = 1, \forall i \in \{1, ..., d\}$. This restriction implies that each dimension of the relation embedding can be represented as $\mathbf{r}_\tau[i] = e^{i\theta_{r,i}}$ and thus corresponds to a rotation in the complex plane.

4.3.1   REPRESENTATIONAL ABILITIES

One way to characterize the various multi-relational decoders is in terms of their ability to represent different logical patterns on relations.

Symmetry and anti-symmetry   For example, many relations are *symmetric*, meaning that

$$(u, \tau, v) \in \mathcal{E} \leftrightarrow (v, \tau, u) \in \mathcal{E}. \tag{4.18}$$

In other cases, we have explicitly *anti-symmetric* relations that satisfy:

$$(u, \tau, v) \in \mathcal{E} \rightarrow (v, \tau, u) \notin \mathcal{E}. \tag{4.19}$$

One important question is whether or not different decoders are capable modeling both symmetric and anti-symmetric relations. DistMult, for example, can only represent symmetric relations, since

$$\text{DEC}(\mathbf{z}_u, \tau, \mathbf{z}_v) = <\mathbf{z}_u, \mathbf{r}_\tau, \mathbf{z}_v>$$
$$= <\mathbf{z}_v, \mathbf{r}_\tau, \mathbf{z}_u>$$
$$= \text{DEC}(\mathbf{z}_v, \tau, \mathbf{z}_u)$$

by definition for that approach. The TransE model on the other hand can only represent anti-symmetric relations, since

$$\text{DEC}(\mathbf{z}_u, \tau, \mathbf{z}_v) = \text{DEC}(\mathbf{z}_v, \tau, \mathbf{z}_u)$$
$$-\|\mathbf{z}_u + \mathbf{r}_\tau - \mathbf{z}_v\| = -\|\mathbf{z}_v + \mathbf{r}_\tau - \mathbf{z}_u\|$$
$$\Rightarrow$$
$$-\mathbf{r}_\tau = \mathbf{r}_\tau$$
$$\Rightarrow$$
$$\mathbf{r}_\tau = 0.$$

Table 4.2: Summary of the ability of some popular multi-relational decoders to encode relational patterns. Adapted from Sun et al. [2019].

Name	Symmetry	Anti-Symmetry	Inversion	Compositionality
RESCAL	✓	✓	✓	✓
TransE	✗	✓	✓	✓
TransX	✓	✓	✗	✗
DistMult	✓	✗	✗	✗
ComplEx	✓	✓	✓	✗
RotatE	✓	✓	✓	✓

Inversion   Related to symmetry is the notion of inversion, where one relation implies the existence of another, with opposite directionality:

$$(u, \tau_1, v) \in \mathcal{E} \leftrightarrow (v, \tau_2, u) \in \mathcal{E}. \tag{4.20}$$

Most decoders are able to represent inverse relations, though again DistMult is unable to model such a pattern.

Compositonality   Last, we can consider whether or not the decoders can encode compositionality between relation representations of the form:

$$(u, \tau_1, y) \in \mathcal{E} \wedge (y, \tau_2, v) \in \mathcal{E} \rightarrow (u, \tau_3, v) \in \mathcal{E}. \tag{4.21}$$

For example, in TransE we can accommodate this by defining $\mathbf{r}_{\tau_3} = \mathbf{r}_{\tau_1} + \mathbf{r}_{\tau_2}$. We can similarly model compositionality in RESCAL by defining $\mathbf{R}_{\tau_3} = \mathbf{R}_{\tau_2} \mathbf{R}_{\tau_1}$.

In general, considering these kinds of relational patterns is useful for comparing the representational capacities of different multi-relational decoders. In practice, we may not expect these patterns to hold exactly, but there may be many relations that exhibit these patterns to some degree (e.g., relations that are symmetric > 90% of the time). Table 4.2 summarizes the ability of the various decoders we discussed to encode these relational patterns.

PART II

Graph Neural Networks

CHAPTER 5

The Graph Neural Network Model

The first part of this book discussed approaches for learning low-dimensional embeddings of the nodes in a graph. The node embedding approaches we discussed used a *shallow* embedding approach to generate representations of nodes, where we simply optimized a unique embedding vector for each node. In this chapter, we turn our focus to more complex encoder models. We will introduce the *graph neural network (GNN)* formalism, which is a general framework for defining deep neural networks on graph data. The key idea is that we want to generate representations of nodes that actually depend on the structure of the graph, as well as any feature information we might have.

The primary challenge in developing complex encoders for graph-structured data is that our usual deep learning toolbox does not apply. For example, convolutional neural networks (CNNs) are well defined only over grid-structured inputs (e.g., images), while recurrent neural networks (RNNs) are well defined only over sequences (e.g., text). To define a deep neural network over general graphs, we need to define a new kind of deep learning architecture.

Permutation invariance and equivariance   One reasonable idea for defining a deep neural network over graphs would be to simply use the adjacency matrix as input to a deep neural network. For example, to generate an embedding of an entire graph we could simply flatten the adjacency matrix and feed the result to a multi-layer perceptron (MLP):

$$\mathbf{z}_G = \mathrm{MLP}(\mathbf{A}[1] \oplus \mathbf{A}[2] \oplus ... \oplus \mathbf{A}[|\mathcal{V}|]), \tag{5.1}$$

where $\mathbf{A}[i] \in \mathbf{R}^{|\mathcal{V}|}$ denotes a row of the adjacency matrix and we use $\oplus$ to denote vector concatenation.

The issue with this approach is that it *depends on the arbitrary ordering of nodes that we used in the adjacency matrix*. In other words, such a model is not *permutation invariant*, and a key desideratum for designing neural networks over graphs is that they should permutation invariant (or equivariant). In mathematical terms, any function f that takes an adjacency matrix $\mathbf{A}$ as input should ideally satisfy one of the two following properties:

$$f(\mathbf{P}\mathbf{A}\mathbf{P}^{\top}) = f(\mathbf{A}) \qquad \text{(Permutation Invariance)} \tag{5.2}$$
$$f(\mathbf{P}\mathbf{A}\mathbf{P}^{\top}) = \mathbf{P}f(\mathbf{A}) \qquad \text{(Permutation Equivariance)}, \tag{5.3}$$

where $\mathbf{P}$ is a permutation matrix. Permutation invariance means that the function does not depend on the arbitrary ordering of the rows/columns in the adjacency matrix, while permutation equivariance means that the output of f is permuted in an consistent way when we permute the adjacency matrix. (The shallow encoders we discussed in Part I are an example of permutation equivariant functions.) Ensuring invariance or equivariance is a key challenge when we are learning over graphs, and we will revisit issues surrounding permutation equivariance and invariance often in the ensuing chapters.

5.1 NEURAL MESSAGE PASSING

The basic GNN model can be motivated in a variety of ways. The same fundamental GNN model has been derived as a generalization of convolutions to non-Euclidean data [Bruna et al., 2014], as a differentiable variant of belief propagation [Dai et al., 2016], as well as by analogy to classic graph isomorphism tests [Hamilton et al., 2017b]. Regardless of the motivation, the defining feature of a GNN is that it uses a form of *neural message passing* in which vector messages are exchanged between nodes and updated using neural networks [Gilmer et al., 2017].

In the rest of this chapter, we will detail the foundations of this neural message passing framework. We will focus on the message passing framework itself and defer discussions of training and optimizing GNN models to Chapter 6. The bulk of this chapter will describe how we can take an input graph $\mathcal{G} = (\mathcal{V}, \mathcal{E})$, along with a set of node features $\mathbf{X} \in \mathbb{R}^{d \times |\mathcal{V}|}$, and use this information to generate node embeddings $\mathbf{z}_u, \forall u \in \mathcal{V}$. However, we will also discuss how the GNN framework can be used to generate embeddings for subgraphs and entire graphs.

5.1.1 OVERVIEW OF THE MESSAGE PASSING FRAMEWORK

During each message-passing iteration in a GNN, a *hidden embedding* $\mathbf{h}_u^{(k)}$ corresponding to each node $u \in \mathcal{V}$ is updated according to information aggregated from u's graph neighborhood $\mathcal{N}(u)$ (Figure 5.1). This message-passing update can be expressed as follows:

$$\mathbf{h}_u^{(k+1)} = \text{UPDATE}^{(k)}\left(\mathbf{h}_u^{(k)}, \text{AGGREGATE}^{(k)}(\{\mathbf{h}_v^{(k)}, \forall v \in \mathcal{N}(u)\})\right) \qquad (5.4)$$

$$= \text{UPDATE}^{(k)}\left(\mathbf{h}_u^{(k)}, \mathbf{m}_{\mathcal{N}(u)}^{(k)}\right), \qquad (5.5)$$

where UPDATE and AGGREGATE are arbitrary differentiable functions (i.e., neural networks) and $\mathbf{m}_{\mathcal{N}(u)}$ is the "message" that is aggregated from u's graph neighborhood $\mathcal{N}(u)$. We use superscripts to distinguish the embeddings and functions at different iterations of message passing.[1]

At each iteration k of the GNN, the AGGREGATE function takes as input the set of embeddings of the nodes in u's graph neighborhood $\mathcal{N}(u)$ and generates a message $\mathbf{m}_{\mathcal{N}(u)}^{(k)}$ based on this aggregated neighborhood information. The update function UPDATE then combines the

---

[1]The different iterations of message passing are also sometimes known as the different "layers" of the GNN.

Figure 5.1: Overview of how a single node aggregates messages from its local neighborhood. The model aggregates messages from A's local graph neighbors (i.e., B, C, and D), and in turn, the messages coming from these neighbors are based on information aggregated from their respective neighborhoods, and so on. This visualization shows a two-layer version of a message-passing model. Notice that the computation graph of the GNN forms a tree structure by unfolding the neighborhood around the target node.

message $\mathbf{m}_{\mathcal{N}(u)}^{(k)}$ with the previous embedding $\mathbf{h}_u^{(k-1)}$ of node u to generate the updated embedding $\mathbf{h}_u^{(k)}$. The initial embeddings at $k = 0$ are set to the input features for all the nodes, i.e., $\mathbf{h}_u^{(0)} = \mathbf{x}_u, \forall u \in \mathcal{V}$. After running K iterations of the GNN message passing, we can use the output of the final layer to define the embeddings for each node, i.e.,

$$\mathbf{z}_u = \mathbf{h}_u^{(K)}, \forall u \in \mathcal{V}. \tag{5.6}$$

Note that since the AGGREGATE function takes a *set* as input, GNNs defined in this way are permutation equivariant by design.

Node features   Note that unlike the shallow embedding methods discussed in Part I of this book, the GNN framework requires that we have node features $\mathbf{x}_u, \forall u \in \mathcal{V}$ as input to the model. In many graphs, we will have rich node features to use (e.g., gene expression features in biological networks or text features in social networks). However, in cases where no node features are available, there are still several options. One option is to use node statistics—such as those introduced in Section 2.1—to define features. Another popular approach is to use *identity features*, where we associate each node with a one-hot indicator feature, which uniquely identifies that node.[a]

[a]Note, however, that the using identity features makes the model transductive and incapable of generalizing to unseen nodes.

5.1.2   MOTIVATIONS AND INTUITIONS

The basic intuition behind the GNN message-passing framework is straightforward: at each iteration, every node aggregates information from its local neighborhood, and as these iterations progress each node embedding contains more and more information from further reaches of the graph. To be precise: after the first iteration ($k = 1$), every node embedding contains information from its 1-hop neighborhood, i.e., every node embedding contains information about the features of its immediate graph neighbors, which can be reached by a path of length 1 in the graph; after the second iteration ($k = 2$) every node embedding contains information from its 2-hop neighborhood; and, in general, after k iterations every node embedding contains information about its k-hop neighborhood.

But what kind of "information" do these node embeddings actually encode? Generally, this information comes in two forms. On the one hand, there is *structural* information about the graph. For example, after k iterations of GNN message passing, the embedding $\mathbf{h}_u^{(k)}$ of node u might encode information about the degrees of all the nodes in u's k-hop neighborhood. This structural information can be useful for many tasks. For instance, when analyzing molecular graphs, we can use degree information to infer atom types and different *structural motifs*, such as benzene rings.

In addition to structural information, the other key kind of information captured by GNN node embedding is *feature-based*. After k iterations of GNN message passing, the embeddings for each node also encode information about all the features in their k-hop neighborhood. This local feature-aggregation behavior of GNNs is analogous to the behavior of the convolutional kernels in CNNs. However, whereas CNNs aggregate feature information from spatially-defined patches in an image, GNNs aggregate information based on local graph neighborhoods. We will explore the connection between GNNs and convolutions in more detail in Chapter 7.

5.1.3   THE BASIC GNN

So far, we have discussed the GNN framework in a relatively abstract fashion as a series of message-passing iterations using UPDATE and AGGREGATE functions (Equation (5.4)). In order to translate the abstract GNN framework defined in Equation (5.4) into something we can implement, we must give concrete instantiations to these UPDATE and AGGREGATE functions. We begin here with the most basic GNN framework, which is a simplification of the original GNN models proposed by Merkwirth and Lengauer [2005] and Scarselli et al. [2009].

The basic GNN message passing is defined as

$$\mathbf{h}_u^{(k)} = \sigma \left(\mathbf{W}_{\text{self}}^{(k)} \mathbf{h}_u^{(k-1)} + \mathbf{W}_{\text{neigh}}^{(k)} \sum_{v \in \mathcal{N}(u)} \mathbf{h}_v^{(k-1)} + \mathbf{b}^{(k)} \right), \tag{5.7}$$

where $\mathbf{W}_{\text{self}}^{(k)}, \mathbf{W}_{\text{neigh}}^{(k)} \in \mathbb{R}^{d^{(k)} \times d^{(k-1)}}$ are trainable parameter matrices and σ denotes an element-wise nonlinearity (e.g., a tanh or ReLU). The bias term $\mathbf{b}^{(k)} \in \mathbb{R}^{d^{(k)}}$ is often omitted for notational simplicity, but including the bias term can be important to achieve strong performance. In this equation—and throughout the remainder of the book—we use superscripts to differentiate parameters, embeddings, and dimensionalities in different layers of the GNN.

The message passing in the basic GNN framework is analogous to a standard multi-layer perceptron (MLP) or Elman-style recurrent neural network, i.e., Elman RNN [Elman, 1990], as it relies on linear operations followed by a single elementwise nonlinearity. We first sum the messages incoming from the neighbors; then, we combine the neighborhood information with the node's previous embedding using a linear combination; and finally, we apply an elementwise nonlinearity.

We can equivalently define the basic GNN through the UPDATE and AGGREGATE functions:

$$m_{\mathcal{N}(u)} = \sum_{v \in \mathcal{N}(u)} h_v, \tag{5.8}$$

$$\text{UPDATE}(h_u, m_{\mathcal{N}(u)}) = \sigma \left(W_{\text{self}} h_u + W_{\text{neigh}} m_{\mathcal{N}(u)} \right), \tag{5.9}$$

where we recall that we use

$$m_{\mathcal{N}(u)} = \text{AGGREGATE}^{(k)}(\{h_v^{(k)}, \forall v \in \mathcal{N}(u)\}) \tag{5.10}$$

as a shorthand to denote the message that has been aggregated from u's graph neighborhood. Note also that we ommitted the superscript denoting the iteration in the above equations, which we will often do for notational brevity.[2]

Node vs. graph-level equations In the description of the basic GNN model above, we defined the core message-passing operations at the node level. We will use this convention for the bulk of this chapter and this book as a whole. However, it is important to note that many GNNs can also be succinctly defined using graph-level equations. In the case of a basic GNN, we can write the graph-level definition of the model as follows:

$$H^{(t)} = \sigma \left(AH^{(k-1)}W_{\text{neigh}}^{(k)} + H^{(k-1)}W_{\text{self}}^{(k)} \right), \tag{5.11}$$

where $H^{(k)} \in \mathbb{R}^{|V| \times d}$ denotes the matrix of node representations at layer t in the GNN (with each node corresponding to a row in the matrix), A is the graph adjacency matrix, and we have omitted the bias term for notational simplicity. While this graph-level representation is not easily applicable to all GNN models—such as the attention-based models we discuss below—it is often more succinct and also highlights how many GNNs can be efficiently implemented using a small number of sparse matrix operations.

---

[2]In general, the parameters W_{self}, W_{neigh}, and b can be shared across the GNN message passing iterations or trained separately for each layer.

5.1.4 MESSAGE PASSING WITH SELF-LOOPS

As a simplification of the neural message passing approach, it is common to add self-loops to the input graph and omit the explicit update step. In this approach we define the message passing simply as

$$\mathbf{h}_u^{(k)} = \text{AGGREGATE}(\{\mathbf{h}_v^{(k-1)}, \forall v \in \mathcal{N}(u) \cup \{u\}\}), \tag{5.12}$$

where now the aggregation is taken over the set $\mathcal{N}(u) \cup \{u\}$, i.e., the node's neighbors as well as the node itself. The benefit of this approach is that we no longer need to define an explicit update function, as the update is implicitly defined through the aggregation method. Simplifying the message passing in this way can often alleviate overfitting, but it also severely limits the expressivity of the GNN, as the information coming the node's neighbors cannot be differentiated from the information from the node itself.

In the case of the basic GNN, adding self-loops is equivalent to sharing parameters between the $\mathbf{W}_{\text{self}}$ and $\mathbf{W}_{\text{neigh}}$ matrices, which gives the following graph-level update:

$$\mathbf{H}^{(t)} = \sigma\left((\mathbf{A} + \mathbf{I})\mathbf{H}^{(t-1)}\mathbf{W}^{(t)}\right). \tag{5.13}$$

In the following chapters we will refer to this as the *self-loop* GNN approach.

5.2 GENERALIZED NEIGHBORHOOD AGGREGATION

The basic GNN model outlined in Equation (5.7) can achieve strong performance, and its theoretical capacity is well-understood (see Chapter 7). However, just like a simple MLP or Elman RNN, the basic GNN can be improved upon and generalized in many ways. Here, we discuss how the AGGREGATE operator can be generalized and improved upon, with the following section (Section 5.3) providing an analogous discussion for the UPDATE operation.

5.2.1 NEIGHBORHOOD NORMALIZATION

The most basic neighborhood aggregation operation (Equation (5.8)) simply takes the sum of the neighbor embeddings. One issue with this approach is that it can be unstable and highly sensitive to node degrees. For instance, suppose node u has $100\times$ as many neighbors as node u' (i.e., a much higher degree), then we would reasonably expect that $\| \sum_{v \in \mathcal{N}(u)} \mathbf{h}_v \| >> \| \sum_{v' \in \mathcal{N}(u')} \mathbf{h}_{v'} \|$ (for any reasonable vector norm $\| \cdot \|$). This drastic difference in magnitude can lead to numerical instabilities as well as difficulties for optimization.

One solution to this problem is to simply normalize the aggregation operation based upon the degrees of the nodes involved. The simplest approach is to just take an average rather than sum:

$$\mathbf{m}_{\mathcal{N}(u)} = \frac{\sum_{v \in \mathcal{N}(u)} \mathbf{h}_v}{|\mathcal{N}(u)|}, \tag{5.14}$$

but researchers have also found success with other normalization factors, such as the following *symmetric normalization* employed by Kipf and Welling [2016a]:

$$\mathbf{m}_{\mathcal{N}(u)} = \sum_{v \in \mathcal{N}(u)} \frac{\mathbf{h}_v}{\sqrt{|\mathcal{N}(u)||\mathcal{N}(v)|}}. \tag{5.15}$$

For example, in a citation graph—the kind of data that Kipf and Welling [2016a] analyzed—information from very high-degree nodes (i.e., papers that are cited many times) may not be very useful for inferring community membership in the graph, since these papers can be cited thousands of times across diverse subfields. Symmetric normalization can also be motivated based on spectral graph theory. In particular, combining the symmetric-normalized aggregation (Equation (5.15)) along with the basic GNN update function (Equation (5.9)) results in a first-order approximation of a spectral graph convolution, and we expand on this connection in Chapter 7.

Graph Convolutional Networks (GCNs)

One of the most popular baseline graph neural network models—the graph convolutional network (GCN)—employs the symmetric-normalized aggregation as well as the self-loop update approach. The GCN model thus defines the message passing function as

$$\mathbf{h}_u^{(k)} = \sigma \left(\mathbf{W}^{(k)} \sum_{v \in \mathcal{N}(u) \cup \{u\}} \frac{\mathbf{h}_v}{\sqrt{|\mathcal{N}(u)||\mathcal{N}(v)|}} \right). \tag{5.16}$$

This approach was first outlined by Kipf and Welling [2016a] and has proved to be one of the most popular and effective baseline GNN architectures.

> **To normalize or not to normalize?** Proper normalization can be essential to achieve stable and strong performance when using a GNN. It is important to note, however, that normalization can also lead to a loss of information. For example, after normalization, it can be hard (or even impossible) to use the learned embeddings to distinguish between nodes of different degrees, and various other structural graph features can be obscured by normalization. In fact, a basic GNN using the normalized aggregation operator in Equation (5.14) is provably less powerful than the basic sum aggregator in Equation (5.8) (see Chapter 7). The use of normalization is thus an application-specific question. Usually, normalization is most helpful in tasks where node feature information is far more useful than structural information, or where there is a very wide range of node degrees that can lead to instabilities during optimization.

5.2.2   SET AGGREGATORS

Neighborhood normalization can be a useful tool to improve GNN performance, but can we do more to improve the AGGREGATE operator? Is there perhaps something more sophisticated than just summing over the neighbor embeddings?

The neighborhood aggregation operation is fundamentally a set function. We are given a set of neighbor embeddings $\{\mathbf{h}_v, \forall v \in \mathcal{N}(u)\}$ and must map this set to a single vector $\mathbf{m}_{\mathcal{N}(u)}$. The fact that $\{\mathbf{h}_v, \forall v \in \mathcal{N}(u)\}$ is a *set* is in fact quite important: there is no natural ordering of a nodes' neighbors, and any aggregation function we define must thus be *permutation invariant*.

Set Pooling

One principled approach to define an aggregation function is based on the theory of permutation invariant neural networks. For example, Zaheer et al. [2017] show that an aggregation function with the following form is a *universal set function approximator*:

$$\mathbf{m}_{\mathcal{N}(u)} = \text{MLP}_\theta \left(\sum_{v \in N(u)} \text{MLP}_\phi(\mathbf{h}_v) \right), \tag{5.17}$$

where as usual we use MLP_θ to denote an arbitrarily deep multi-layer perceptron parameterized by some trainable parameters θ. In other words, the theoretical results in Zaheer et al. [2017] show that any permutation-invariant function that maps a set of embeddings to a single embedding can be approximated to an arbitrary accuracy by a model following Equation (5.17).

Note that the theory presented in Zaheer et al. [2017] employs a sum of the embeddings after applying the first MLP (as in Equation (5.17)). However, it is possible to replace the sum with an alternative reduction function, such as an element-wise maximum or minimum, as in Qi et al. [2017], and it is also common to combine models based on Equation (5.17) with the normalization approaches discussed in Section 5.2.1, as in the `GraphSAGE-pool` approach [Hamilton et al., 2017b].

Set pooling approaches based on Equation (5.17) often lead to small increases in performance, though they also introduce an increased risk of overfitting, depending on the depth of the MLPs used. If set pooling is used, it is common to use MLPs that have only a single hidden layer, since these models are sufficient to satisfy the theory, but are not so overparameterized so as to risk catastrophic overfitting.

Janossy Pooling

Set pooling approaches to neighborhood aggregation essentially just add extra layers of MLPs on top of the more basic aggregation architectures discussed in Section 5.1.3. This idea is simple, but is known to increase the theoretical capacity of GNNs. However, there is another alternative approach, termed *Janossy pooling*, that is also provably more powerful than simply taking a sum or mean of the neighbor embeddings [Murphy et al., 2018].

Recall that the challenge of neighborhood aggregation is that we must use a *permutation-invariant* function, since there is no natural ordering of a node's neighbors. In the set pooling approach (Equation (5.17)), we achieved this permutation invariance by relying on a sum, mean, or element-wise max to reduce the set of embeddings to a single vector. We made the model more powerful by combining this reduction with feed-forward neural networks (i.e., MLPs). Janossy pooling employs a different approach entirely: instead of using a permutation-invariant reduction (e.g., a sum or mean), we apply a *permutation-sensitive function* and average the result over many possible permutations.

Let $\pi_i \in \Pi$ denote a permutation function that maps the set $\{\mathbf{h}_v, \forall v \in \mathcal{N}(u)\}$ to a specific sequence $(\mathbf{h}_{v_1}, \mathbf{h}_{v_2}, ..., \mathbf{h}_{v_{|\mathcal{N}(u)|}})_{\pi_i}$. In other words, π_i takes the unordered set of neighbor embeddings and places these embeddings in a sequence based on some arbitrary ordering. The Janossy pooling approach then performs neighborhood aggregation by

$$\mathbf{m}_{\mathcal{N}(u)} = \text{MLP}_\theta \left(\frac{1}{|\Pi|} \sum_{\pi \in \Pi} \rho_\phi \left(\mathbf{h}_{v_1}, \mathbf{h}_{v_2}, \mathbf{h}_{v_{|\mathcal{N}(u)|}} \right)_{\pi_i} \right), \tag{5.18}$$

where Π denotes a set of permutations and ρ_ϕ is a permutation-sensitive function, e.g., a neural network that operates on sequences. In practice, ρ_ϕ is usually defined to be an LSTM [Hochreiter and Schmidhuber, 1997], since LSTMs are known to be a powerful neural network architecture for sequences.

If the set of permutations Π in Equation (5.18) is equal to all possible permutations, then the aggregator in Equation (5.18) is also a universal function approximator for sets, like Equation (5.17). However, summing over all possible permutations is generally intractable. Thus, in practice, Janossy pooling employs one of two approaches:

1. Sample a random subset of possible permutations during each application of the aggregator, and only sum over that random subset.

2. Employ a *canonical* ordering of the nodes in the neighborhood set, e.g., order the nodes in descending order according to their degree, with ties broken randomly.

Murphy et al. [2018] include a detailed discussion and empirical comparison of these two approaches, as well as other approximation techniques (e.g., truncating the length of sequence), and their results indicate that Janossy-style pooling can improve upon set pooling in a number of synthetic evaluation setups.

5.2.3 NEIGHBORHOOD ATTENTION

In addition to more general forms of set aggregation, a popular strategy for improving the aggregation layer in GNNs is to apply *attention* [Bahdanau et al., 2015]. The basic idea is to assign an attention weight or importance to each neighbor, which is used to weigh this neighbor's influence during the aggregation step. The first GNN model to apply this style of attention was Veličković et al. [2018]'s Graph Attention Network (GAT), which uses attention weights to define a weighted sum of the neighbors:

$$\mathbf{m}_{\mathcal{N}(u)} = \sum_{v \in \mathcal{N}(u)} \alpha_{u,v} \mathbf{h}_v, \tag{5.19}$$

where $\alpha_{u,v}$ denotes the attention on neighbor $v \in \mathcal{N}(u)$ when we are aggregating information at node u. In the original GAT paper, the attention weights are defined as

$$\alpha_{u,v} = \frac{\exp\left(\mathbf{a}^\top [\mathbf{W}\mathbf{h}_u \oplus \mathbf{W}\mathbf{h}_v]\right)}{\sum_{v' \in \mathcal{N}(u)} \exp\left(\mathbf{a}^\top [\mathbf{W}\mathbf{h}_u \oplus \mathbf{W}\mathbf{h}_{v'}]\right)}, \tag{5.20}$$

where $\mathbf{a}$ is a trainable attention vector, $\mathbf{W}$ is a trainable matrix, and $\oplus$ denotes the concatenation operation.

The GAT-style attention computation is known to work well with graph data. However, in principle any standard attention model from the deep learning literature at large can be used [Bahdanau et al., 2015]. Popular variants of attention include the bilinear attention model

$$\alpha_{u,v} = \frac{\exp\left(\mathbf{h}_u^\top \mathbf{W}\mathbf{h}_v\right)}{\sum_{v' \in \mathcal{N}(u)} \exp\left(\mathbf{h}_u^\top \mathbf{W}\mathbf{h}_{v'}\right)}, \tag{5.21}$$

as well as variations of attention layers using MLPs, e.g.,

$$\alpha_{u,v} = \frac{\exp\left(\mathrm{MLP}(\mathbf{h}_u, \mathbf{h}_v)\right)}{\sum_{v' \in \mathcal{N}(u)} \exp\left(\mathrm{MLP}(\mathbf{h}_u, \mathbf{h}_{v'})\right)}, \tag{5.22}$$

where the MLP is restricted to a scalar output.

In addition, while it is less common in the GNN literature, it is also possible to add multiple attention "heads," in the style of the popular *transformer* architecture [Vaswani et al., 2017]. In this approach, one computes K distinct attention weights $\alpha_{u,v,k}$, using independently parameterized attention layers. The messages aggregated using the different attention weights are then transformed and combined in the aggregation step, usually with a linear projection followed by a concatenation operation, e.g.,

$$\mathbf{m}_{\mathcal{N}(u)} = [\mathbf{a}_1 \oplus \mathbf{a}_2 \oplus \dots \oplus \mathbf{a}_K] \tag{5.23}$$
$$\mathbf{a}_k = \mathbf{W}_i \sum_{v \in \mathcal{N}(u)} \alpha_{u,v,k} \mathbf{h}_v, \tag{5.24}$$

where the attention weights $\alpha_{u,v,k}$ for each of the K attention heads can be computed using any of the above attention mechanisms.

Graph attention and transformers   GNN models with multi-headed attention (Equation (5.23)) are closely related to the transformer architecture [Vaswani et al., 2017]. Transformers are a popular architecture for both natural language processing (NLP) and computer vision, and—in the case of NLP—they have been an important driver behind large state-of-the-art NLP systems, such as BERT [Devlin et al., 2018] and XLNet [Yang et al., 2019]. The basic idea behind transformers is to define neural network layers entirely based on the attention operation. At each layer in a transformer, a new hidden representation is generated for every position in the input data (e.g., every word in a sentence) by using multiple attention heads to compute attention weights between all pairs of positions in the input, which are then aggregated with weighted sums based on these attention weights (in a manner analogous to Equation (5.23)). In fact, the basic transformer layer is exactly equivalent to a GNN layer using multi-headed attention (i.e., Equation (5.23)) if we assume that the GNN receives a fully connected graph as input.

This connection between GNNs and transformers has been exploited in numerous works. For example, one implementation strategy for designing GNNs is to simply start with a transformer model and then apply a binary adjacency mask on the attention layer to ensure that information is only aggregated between nodes that are actually connected in the graph. This style of GNN implementation can benefit from the numerous well-engineered libraries for transformer architectures that exist. However, a downside of this approach is that transformers must compute the pairwise attention between all positions/nodes in the input, which leads to a quadratic $O(|\mathcal{V}|^2)$ time complexity to aggregate messages for all nodes in the graph, compared to a $O(|\mathcal{V}||\mathcal{E}|)$ time complexity for a more standard GNN implementation.

Adding attention is a useful strategy for increasing the representational capacity of a GNN model, especially in cases where you have prior knowledge to indicate that some neighbors might be more informative than others. For example, consider the case of classifying papers into topical categories based on citation networks. Often there are papers that span topical boundaries, or that are highly cited across various different fields. Ideally, an attention-based GNN would learn to ignore these papers in the neural message passing, as such promiscuous neighbors would likely be uninformative when trying to identify the topical category of a particular node. In Chapter 7, we will discuss how attention can influence the inductive bias of GNNs from a signal processing perspective.

5.3   GENERALIZED UPDATE METHODS

The AGGREGATE operator in GNN models has generally received the most attention from researchers—in terms of proposing novel architectures and variations. This was especially the case after the introduction of the GraphSAGE framework, which introduced the idea of generalized neighborhood aggregation [Hamilton et al., 2017b]. However, GNN message passing involves two key steps: aggregation and updating, and in many ways the UPDATE operator plays an equally important role in defining the power and inductive bias of the GNN model.

So far, we have seen the basic GNN approach—where the update operation involves a linear combination of the node's current embedding with the message from its neighbors—as well as the self-loop approach, which simply involves adding a self-loop to the graph before performing neighborhood aggregation. In this section, we turn our attention to more diverse generalizations of the UPDATE operator.

Over-smoothing and neighborhood influence   One common issue with GNNs—which generalized update methods can help to address—is known as *over-smoothing*. The essential idea of over-smoothing is that after several iterations of GNN message passing, the representations for all the nodes in the graph can become very similar to one another. This tendency is especially common in basic GNN models and models that employ the self-loop update approach. Over-smoothing is problematic because it makes it impossible to build deeper GNN models—which leverage longer-term dependencies in the graph—since these deep GNN models tend to just generate over-smoothed embeddings.

This issue of over-smoothing in GNNs can be formalized by defining the influence of each node's input feature $\mathbf{h}_u^{(0)} = \mathbf{x}_u$ on the final layer embedding of all the other nodes in the graph, i.e, $\mathbf{h}_v^{(K)}, \forall v \in \mathcal{V}$. In particular, for any pair of nodes u and v we can quantify the influence of node u on node v in the GNN by examining the magnitude of the corresponding Jacobian matrix [Xu et al., 2018]:

$$I_K(u,v) = \mathbf{1}^\top \left(\frac{\partial \mathbf{h}_v^{(K)}}{\partial \mathbf{h}_u^{(0)}} \right) \mathbf{1}, \tag{5.25}$$

where $\mathbf{1}$ is a vector of all ones. The $I_K(u,v)$ value uses the sum of the entries in the Jacobian matrix $\frac{\partial \mathbf{h}_v^{(K)}}{\partial \mathbf{h}_u^{(0)}}$ as a measure of how much the initial embedding of node u influences the final embedding of node v in the GNN.

Given this definition of influence, Xu et al. [2018] prove the following:

Theorem 5.1   *For any GNN model using a self-loop update approach and an aggregation function of the form*

$$\text{AGGREGATE}(\{\mathbf{h}_v, \forall v \in \mathcal{N}(u) \cup \{u\}\}) = \frac{1}{f_n(|\mathcal{N}(u) \cup \{u\}|)} \sum_{v \in \mathcal{N}(u) \cup \{u\}} \mathbf{h}_v, \tag{5.26}$$

where $f : \mathbb{R}^+ \to \mathbb{R}^+$ is an arbitrary differentiable normalization function, we have that

$$I_K(u,v) \propto p_{\mathcal{G},K}(u|v), \tag{5.27}$$

where $p_{\mathcal{G},K}(u|v)$ denotes the probability of visiting node v on a length-K random walk starting from node u.

This theorem is a direct consequence of Theorem 1 in Xu et al. [2018]. It states that when we are using a K-layer GCN-style model, the influence of node u and node v is proportional the probability of reaching node v on a K-step random walk starting from node u. An important consequence of this, however, is that as $K \to \infty$ the influence of every node approaches the stationary distribution of random walks over the graph, meaning that local neighborhood information is lost. Moreover, in many real-world graphs—which contain high-degree nodes and resemble so-called "expander" graphs—it only takes $k = O(\log(|\mathcal{V}|))$ steps for the random walk starting from any node to converge to an almost-uniform distribution [Hoory et al., 2006].

Theorem 5.1 applies directly to models using a self-loop update approach, but the result can also be extended in asymptotic sense for the basic GNN update (i.e., Equation (5.9)) as long as $\|\mathbf{W}_{\text{self}}^{(k)}\| < \|\mathbf{W}_{\text{neigh}}^{(k)}\|$ at each layer k. Thus, when using simple GNN models—and especially those with the self-loop update approach—building deeper models can actually hurt performance. As more layers are added we lose information about local neighborhood structures and our learned embeddings become over-smoothed, approaching an almost-uniform distribution.

5.3.1   CONCATENATION AND SKIP-CONNECTIONS

As discussed above, over-smoothing is a core issue in GNNs. Over-smoothing occurs when node-specific information becomes "washed out" or "lost" after several iterations of GNN message passing. Intuitively, we can expect over-smoothing in cases where the information being aggregated from the node neighbors during message passing begins to dominate the updated node representations. In these cases, the updated node representations (i.e., the $\mathbf{h}_u^{(k+1)}$ vectors) will depend too strongly on the incoming message aggregated from the neighbors (i.e., the $\mathbf{m}_{\mathcal{N}(u)}$ vectors) at the expense of the node representations from the previous layers (i.e., the $\mathbf{h}_u^{(k)}$ vectors). A natural way to alleviate this issue is to use vector concatenations or *skip connections*, which try to directly preserve information from previous rounds of message passing during the update step.

These concatenation and skip-connection methods can be used in conjunction with most other GNN update approaches. Thus, for the sake of generality, we will use UPDATE$_{\text{base}}$ to denote the base update function that we are building upon (e.g., we can assume that UPDATE$_{\text{base}}$ is given by Equation (5.9)), and we will define various skip-connection updates on top of this base function.

One of the simplest skip connection updates employs a concatenation to preserve more node-level information during message passing:

$$\text{UPDATE}_{\text{concat}}(\mathbf{h}_u, \mathbf{m}_{\mathcal{N}(u)}) = [\text{UPDATE}_{\text{base}}(\mathbf{h}_u, \mathbf{m}_{\mathcal{N}(u)}) \oplus \mathbf{h}_u], \tag{5.28}$$

where we simply concatenate the output of the base update function with the node's previous-layer representation. Again, the key intuition here is that we encourage the model to disentangle information during message passing—separating the information coming from the neighbors (i.e., $\mathbf{m}_{\mathcal{N}(u)}$) from the current representation of each node (i.e., $\mathbf{h}_u$).

The concatenation-based skip connection was proposed in the `GraphSAGE` framework, which was one of the first works to highlight the possible benefits of these kinds of modifications to the update function [Hamilton et al., 2017a]. However, in addition to concatenation, we can also employ other forms of skip-connections, such as the linear interpolation method proposed by Pham et al. [2017]:

$$\text{UPDATE}_{\text{interpolate}}(\mathbf{h}_u, \mathbf{m}_{\mathcal{N}(u)}) = \boldsymbol{\alpha}_1 \circ \text{UPDATE}_{\text{base}}(\mathbf{h}_u, \mathbf{m}_{\mathcal{N}(u)}) + \boldsymbol{\alpha}_2 \odot \mathbf{h}_u, \tag{5.29}$$

where $\boldsymbol{\alpha}_1, \boldsymbol{\alpha}_2 \in [0, 1]^d$ are gating vectors with $\boldsymbol{\alpha}_2 = 1 - \boldsymbol{\alpha}_1$ and $\circ$ denotes elementwise multiplication. In this approach, the final updated representation is a linear interpolation between the previous representation and the representation that was updated based on the neighborhood information. The gating parameters $\boldsymbol{\alpha}_1$ can be learned jointly with the model in a variety of ways. For example, Pham et al. [2017] generate $\boldsymbol{\alpha}_1$ as the output of a separate single-layer GNN, which takes the current hidden-layer representations as features. However, other simpler approaches could also be employed, such as simply directly learning $\boldsymbol{\alpha}_1$ parameters for each message passing layer or using an MLP on the current node representations to generate these gating parameters.

In general, these concatenation and residual connections are simple strategies that can help to alleviate the over-smoothing issue in GNNs, while also improving the numerical stability of optimization. Indeed, similar to the utility of residual connections in CNNs [He et al., 2016], applying these approaches to GNNs can facilitate the training of much deeper models. In practice, these techniques tend to be most useful for node classification tasks with moderately deep GNNs (e.g., 2–5 layers), and they excel on tasks that exhibit homophily, i.e., where the prediction for each node is strongly related to the features of its local neighborhood.

5.3.2   GATED UPDATES

In the previous section we discussed skip-connection and residual connection approaches that bear strong analogy to techniques used in computer vision to build deeper CNN architectures. In a parallel line of work, researchers have also drawn inspiration from the gating methods used to improve the stability and learning ability of RNNs. In particular, one way to view the GNN message passing algorithm is that the aggregation function is receiving an *observation* from the neighbors, which is then used to update the *hidden state* of each node. In this view, we can directly apply methods used to update the hidden state of RNN architectures based on observations. For instance, one of the earliest GNN architectures [Li et al., 2015] defines the update function as

$$\mathbf{h}_u^{(k)} = \text{GRU}(\mathbf{h}_u^{(k-1)}, \mathbf{m}_{\mathcal{N}(u)}^{(k)}), \tag{5.30}$$

where GRU denotes the update equation of the GRU cell [Cho et al., 2014]. Other approaches have employed updates based on the LSTM architecture [Selsam et al., 2019].

In general, any update function defined for RNNs can be employed in the context of GNNs. We simply replace the hidden state argument of the RNN update function (usually denoted $\mathbf{h}^{(t)}$) with the node's hidden state, and we replace the observation vector (usually denoted $\mathbf{x}^{(t)}$) with the message aggregated from the local neighborhood. Importantly, the parameters of this RNN-style update are always shared across nodes (i.e., we use the same LSTM or GRU cell to update each node). In practice, researchers usually share the parameters of the update function across the message-passing layers of the GNN as well.

These gated updates are very effective at facilitating deep GNN architectures (e.g., more than 10 layers) and preventing over-smoothing. Generally, they are most useful for GNN applications where the prediction task requires complex reasoning over the global structure of the graph, such as applications for program analysis [Li et al., 2015] or combinatorial optimization [Selsam et al., 2019].

5.3.3 JUMPING KNOWLEDGE CONNECTIONS

In the preceding sections, we have been implicitly assuming that we are using the output of the final layer of the GNN. In other words, we have been assuming that the node representations $\mathbf{z}_u$ that we use in a downstream task are equal to final layer node embeddings in the GNN:

$$\mathbf{z}_u = \mathbf{h}_u^{(K)}, \forall u \in \mathcal{V}. \tag{5.31}$$

This assumption is made by many GNN approaches, and the limitations of this strategy motivated much of the need for residual and gated updates to limit over-smoothing.

However, a complimentary strategy to improve the quality of the final node representations is to simply leverage the representations *at each layer of message passing*, rather than only using the final layer output. In this approach we define the final node representations $\mathbf{z}_u$ as

$$\mathbf{z}_u = f_{\text{JK}}(\mathbf{h}_u^{(0)} \oplus \mathbf{h}_u^{(1)} \oplus ... \oplus \mathbf{h}_u^{(K)}), \tag{5.32}$$

where f_{JK} is an arbitrary differentiable function. This strategy is known as adding *jumping knowledge (JK) connections* and was first proposed and analyzed by Xu et al. [2018]. In many applications the function f_{JK} can simply be defined as the identity function, meaning that we just concatenate the node embeddings from each layer, but Xu et al. [2018] also explore other options such as max-pooling approaches and LSTM attention layers. This approach often leads to consistent improvements across a wide variety of tasks and is a generally useful strategy to employ.

5.4   EDGE FEATURES AND MULTI-RELATIONAL GNNS

So far our discussion of GNNs and neural message passing has implicitly assumed that we have simple graphs. However, there are many applications where the graphs in question are multi-relational or otherwise heterogenous (e.g., knowledge graphs). In this section, we review some of the most popular strategies that have been developed to accommodate such data.

5.4.1   RELATIONAL GRAPH NEURAL NETWORKS

The first approach proposed to address this problem is commonly known as the *Relational Graph Convolutional Network (RGCN)* approach [Schlichtkrull et al., 2017]. In this approach we augment the aggregation function to accommodate multiple relation types by specifying a separate transformation matrix per relation type:

$$\mathbf{m}_{\mathcal{N}(u)} = \sum_{\tau \in \mathcal{R}} \sum_{v \in \mathcal{N}_\tau(u)} \frac{\mathbf{W}_\tau \mathbf{h}_v}{f_n(\mathcal{N}(u), \mathcal{N}(v))}, \tag{5.33}$$

where f_n is a normalization function that can depend on both the neighborhood of the node u as well as the neighbor v being aggregated over. Schlichtkrull et al. [2017] discuss several normalization strategies to define f_n that are analogous to those discussed in Section 5.2.1. Overall, the multi-relational aggregation in RGCN is thus analogous to the basic a GNN approach with normalization, but we separately aggregate information across different edge types.

Parameter Sharing
One drawback of the naive RGCN approach is the drastic increase in the number of parameters, as now we have one trainable matrix per relation type. In certain applications—such as applications on knowledge graphs with many distinct relations—this increase in parameters can lead to overfitting and slow learning. Schlichtkrull et al. [2017] propose a scheme to combat this issue by parameter sharing with basis matrices, where

$$\mathbf{W}_\tau = \sum_{i=1}^{b} \alpha_{i,\tau} \mathbf{B}_i. \tag{5.34}$$

In this basis matrix approach, all the relation matrices are defined as linear combinations of b basis matrices $\mathbf{B}_1, ..., \mathbf{B}_b$, and the only relation-specific parameters are the b combination weights $\alpha_{1,\tau}, ..., \alpha_{b,\tau}$ for each relation τ. In this basis sharing approach, we can thus rewrite the full aggregation function as

$$\mathbf{m}_{\mathcal{N}(u)} = \sum_{\tau \in \mathcal{R}} \sum_{v \in \mathcal{N}_\tau(u)} \frac{\boldsymbol{\alpha}_\tau \times_1 \mathcal{B} \times_2 \mathbf{h}_v}{f_n(\mathcal{N}(u), \mathcal{N}(v))}, \tag{5.35}$$

where $\mathcal{B} = (\mathbf{B}_1, ..., \mathbf{B}_b)$ is a tensor formed by stacking the basis matrices, $\boldsymbol{\alpha}_\tau = (\alpha_{1,\tau}, ..., \alpha_{b,\tau})$ is a vector containing the basis combination weights for relation τ, and $\times_i$ denotes a tensor product along mode i. Thus, an alternative view of the parameter sharing RGCN approach is that we are learning an embedding for each relation, as well a tensor that is shared across all relations.

Extensions and Variations
The RGCN architecture can be extended in many ways, and in general, we refer to approaches that define separate aggregation matrices per relation as *relational graph neural networks*. For example, a variation of this approach without parameter sharing is deployed by Zitnik et al. [2018] to model a multi-relational dataset relating drugs, diseases and proteins, and a similar strategy is leveraged by Marcheggiani and Titov [2017] to analyze linguistic dependency graphs. Other works have found success combining the RGCN-style aggregation with attention [Teru et al., 2020].

5.4.2   ATTENTION AND FEATURE CONCATENATION
The relational GNN approach, where we define a separate aggregation parameter per relation, is applicable for multi-relational graphs and cases where we have discrete edge features. To accommodate cases where we have more general forms of edge features, we can leverage these features in attention or by concatenating this information with the neighbor embeddings during message passing. For example, given any base aggregation approach AGGREGATE$_{base}$ one simple strategy to leverage edge features is to define a new aggregation function as

$$\mathbf{m}_{\mathcal{N}(u)} = \text{AGGREGATE}_{base}(\{\mathbf{h}_v \oplus \mathbf{e}_{(u,\tau,v)}, \forall v \in \mathcal{N}(u)\}), \tag{5.36}$$

where $\mathbf{e}_{(u,\tau,v)}$ denotes an arbitrary vector-valued feature for the edge (u, τ, v). This approach is simple and general, and has seen recent success with attention-based approaches as the base aggregation function [Sinha et al., 2019].

5.5   GRAPH POOLING
The neural message passing approach produces a set of *node* embeddings, but what if we want to make predictions at the *graph* level? In other words, we have been assuming that the goal is to learn node representations $\mathbf{z}_u, \forall u \in \mathcal{V}$, but what if we to learn an embedding $\mathbf{z}_{\mathcal{G}}$ for the entire graph $\mathcal{G}$? This task is often referred to as *graph pooling*, since our goal is to pool together the node embeddings in order to learn an embedding of the entire graph.

Set Pooling Approaches
Similar to the AGGREGATE operator, the task of graph pooling can be viewed as a problem of learning over sets. We want to design a pooling function f_p, which maps a set of node embeddings $\{\mathbf{z}_1, ..., \mathbf{z}_{|V|}\}$ to an embedding $\mathbf{z}_{\mathcal{G}}$ that represents the full graph. Indeed, any of the approaches discussed in Section 5.2.2 for learning over sets of neighbor embeddings can also be employed for pooling at the graph level.

In practice, there are two approaches that are most commonly applied for learning graph-level embeddings via set pooling. The first approach is simply to take a sum (or mean) of the node embeddings:

$$\mathbf{z}_\mathcal{G} = \frac{\sum_{v \in \mathcal{V}} \mathbf{z}_c}{f_n(|\mathcal{V}|)}, \tag{5.37}$$

where f_n is some normalizing function (e.g., the identity function). While quite simple, pooling based on the sum or mean of the node embeddings is often sufficient for applications involving small graphs.

The second popular set-based approach uses a combination of LSTMs and attention to pool the node embeddings, in a manner inspired by the work of Vinyals et al. [2015]. In this pooling approach, we iterate a series of attention-based aggregations defined by the following set of equations, which are iterated for $t = 1, ..., T$ steps:

$$\mathbf{q}_i = \text{LSTM}(\mathbf{o}_{t-1}, \mathbf{q}_{t-1}), \tag{5.38}$$
$$e_{v,t} = f_a(\mathbf{z}_v, \mathbf{q}_t), \forall v \in \mathcal{V}, \tag{5.39}$$
$$a_{v,t} = \frac{\exp(e_{v,i})}{\sum_{u \in \mathcal{V}} e_{u,t}}, \forall v \in \mathcal{V}, \tag{5.40}$$
$$\mathbf{o}_t = \sum_{v \in \mathcal{V}} a_{v,t} \mathbf{z}_v. \tag{5.41}$$

In the above equations, the $\mathbf{q}_t$ vector represents a *query vector* for the attention at each iteration t. In Equation (5.39), the query vector is used to compute an attention score over each node using an attention function $f_a : \mathbb{R}^d \times \mathbb{R}^d \rightarrow \mathbb{R}$ (e.g., a dot product), and this attention score is then normalized in Equation (5.40). Finally, in Equation (5.41) a weighted sum of the node embeddings is computed based on the attention weights, and this weighted sum is used to update the query vector using an LSTM update (Equation (5.38)). Generally, the $\mathbf{q}_0$ and $\mathbf{o}_0$ vectors are initialized with all-zero values, and after iterating Equations (5.38)–(5.41) for T iterations, an embedding for the full graph is computed as

$$\mathbf{z}_\mathcal{G} = \mathbf{o}_1 \oplus \mathbf{o}_2 \oplus ... \oplus \mathbf{o}_T. \tag{5.42}$$

This approach represents a sophisticated architecture for attention-based pooling over a set, and it has become a popular pooling method in many graph-level classification tasks.

Graph Coarsening Approaches

One limitation of the set pooling approaches is that they do not exploit the structure of the graph. While it is reasonable to consider the task of graph pooling as simply a set learning problem, there can also be benefits from exploiting the graph topology at the pooling stage. One popular strategy to accomplish this is to perform graph *clustering or coarsening* as a means to pool the node representations.

In these style of approaches, we assume that we have some clustering function

$$\mathbf{f}_c \to \mathcal{G} \times \mathbb{R}^{|V| \times d} \to \mathbb{R}^{+|V| \times c}, \tag{5.43}$$

which maps all the nodes in the graph to an assignment over c clusters. In particular, we presume that this function outputs an assignment matrix $\mathbf{S} = f_c(\mathcal{G}, \mathbf{Z})$, where $\mathbf{S}[u, i] \in \mathbb{R}^+$ denotes the strength of the association between node u and cluster i. One simple example of an f_c function would be spectral clustering approach described in Chapter 1, where the cluster assignment is based on the spectral decomposition of the graph adjacency matrix. In a more complex definition of f_c, one can actually employ another GNN to predict cluster assignments [Ying et al., 2018b].

Regardless of the approach used to generate the cluster assignment matrix $\mathbf{S}$, the key idea of graph coarsening approaches is that we then use this matrix to *coarsen* the graph. In particular, we use the assignment matrix $\mathbf{S}$ to compute a new coarsened adjacency matrix

$$\mathbf{A}^{\text{new}} = \mathbf{S}^\top \mathbf{A} \mathbf{S} \in \mathbb{R}^{+c \times c} \tag{5.44}$$

and a new set of node features

$$\mathbf{X}^{\text{new}} = \mathbf{S}^\top \mathbf{X} \in \mathbb{R}^{c \times d}. \tag{5.45}$$

Thus, this new adjacency matrix now represents the strength of association (i.e., the edges) between the clusters in the graph, and the new feature matrix represents the aggregated embeddings for all the nodes assigned to each cluster. We can then run a GNN on this coarsened graph and repeat the entire coarsening process for a number of iterations, where the size of the graph is decreased at each step. The final representation of the graph is then computed by a set pooling over the embeddings of the nodes in a sufficiently coarsened graph.

This coarsening based approach is inspired by the pooling approaches used in CNNs, and it relies on the intuition that we can build hierarchical GNNs that operate on different granularities of the input graph. In practice, these coarsening approaches can lead to strong performance, but they can also be unstable and difficult to train. For example, in order to have the entire learning process be end-to-end differentiable the clustering functions f_c must be differentiable, which rules out most off-the-shelf clustering algorithms such as spectral clustering. There are also approaches that coarsen the graph by selecting a set of nodes to remove rather than pooling all nodes into clusters, which can lead to benefits in terms of computational complexity, speed, and downstream performance [Cangea et al., 2018, Gao and Ji, 2019].

5.6   GENERALIZED MESSAGE PASSING

The presentation in this chapter so far has focused on the most popular style of GNN message passing, which operates largely at the node level. However, the GNN message passing approach can also be generalized to leverage edge and graph-level information at each stage of message passing. For example, in the more general approach proposed by Battaglia et al. [2018], we define each iteration of message passing according to the following equations:

$$h_{(u,v)}^{(k)} = \text{UPDATE}_{\text{edge}}\left(h_{(u,v)}^{(k-1)}, h_u^{(k-1)}, h_v^{(k-1)}, h_{\mathcal{G}}^{(k-1)}\right) \tag{5.46}$$

$$m_{\mathcal{N}(u)} = \text{AGGREGATE}_{\text{node}}\left(\{h_{(u,v)}^{(k)}, \forall v \in \mathcal{N}(u)\}\right) \tag{5.47}$$

$$h_u^{(k)} = \text{UPDATE}_{\text{node}}\left(h_u^{(k-1)}, m_{\mathcal{N}(u)}, h_{\mathcal{G}}^{(k-1)}\right) \tag{5.48}$$

$$h_{\mathcal{G}}^{(k)} = \text{UPDATE}_{\text{graph}}\left(h_{\mathcal{G}}^{(k-1)}, \{h_u^{(k)}, \forall u \in \mathcal{V}\}, \{h_{(u,v)}^{(k)}, \forall (u,v) \in \mathcal{E}\}\right). \tag{5.49}$$

The important innovation in this generalized message passing framework is that, during message passing, we generate hidden embeddings $h_{(u,v)}^{(k)}$ for each edge in the graph, as well as an embedding $h_{\mathcal{G}}^{(k)}$ corresponding to the entire graph. This allows the message passing model to easily integrate edge and graph-level features, and recent work has also shown this generalized message passing approach to have benefits compared to a standard GNN in terms of logical expressiveness [Barceló et al., 2020]. Generating embeddings for edges and the entire graph during message passing also makes it trivial to define loss functions based on graph or edge-level classification tasks.

In terms of the message-passing operations in this generalized message-passing framework, we first update the edge embeddings based on the embeddings of their incident nodes (Equation (5.46)). Next, we update the node embeddings by aggregating the edge embeddings for all their incident edges (Equations (5.47) and (5.48)). The graph embedding is used in the update equation for both node and edge representations, and the graph-level embedding itself is updated by aggregating over all the node and edge embeddings at the end of each iteration (Equation (5.49)). All of the individual update and aggregation operations in such a generalized message-passing framework can be implemented using the techniques discussed in this chapter (e.g., using a pooling method to compute the graph-level update).

CHAPTER 6

Graph Neural Networks in Practice

In Chapter 5, we introduced a number of GNN architectures. However, we did not discuss how these architectures are optimized and what kinds of loss functions and regularization are generally used. In this chapter, we will turn our attention to some of these practical aspects of GNNs. We will discuss some representative applications and how GNNs are generally optimized in practice, including a discussion of unsupervised pre-training methods that can be particularly effective. We will also introduce common techniques used to regularize and improve the efficiency of GNNs.

6.1 APPLICATIONS AND LOSS FUNCTIONS

In the vast majority of current applications, GNNs are used for one of three tasks: node classification, graph classification, or relation prediction. As discussed in Chapter 1, these tasks reflect a large number of real-world applications, such as predicting whether a user is a bot in a social network (node classification), property prediction based on molecular graph structures (graph classification), and content recommendation in online platforms (relation prediction). In this section, we briefly describe how these tasks translate into concrete loss functions for GNNs, and we also discuss how GNNs can be pre-trained in an unsupervised manner to improve performance on these downstream tasks.

In the following discussions, we will use $z_u \in \mathbb{R}^d$ to denote the node embedding output by the final layer of a GNN, and we will use $z_G \in \mathbb{R}^d$ to denote a graph-level embedding output by a pooling function. Any of the GNN approaches discussed in Chapter 5 could, in principle, be used to generate these embeddings. In general, we will define loss functions on the z_u and z_G embeddings, and we will assume that the gradient of the loss is backpropagated through the parameters of the GNN using stochastic gradient descent or one of its variants [Rumelhart et al., 1986].

6.1.1 GNNS FOR NODE CLASSIFICATION

Node classification is one of the most popular benchmark tasks for GNNs. For instance, during the years 2017–2019—when GNN methods were beginning to gain prominence across machine learning—research on GNNs was dominated by the Cora, Citeseer, and Pubmed citation network benchmarks, which were popularized by Kipf and Welling [2016a]. These baselines involved classifying the category or topic of scientific papers based on their position within a citation network, with language-based node features (e.g., word vectors) and only a very small number of positive examples given per each class (usually less than 10% of the nodes).

The standard way to apply GNNs to such a node classification task is to train GNNs in a fully-supervised manner, where we define the loss using a softmax classification function and negative log-likelihood loss:

$$\mathcal{L} = \sum_{u \in \mathcal{V}_{\text{train}}} -\log(\text{softmax}(\mathbf{z}_u, \mathbf{y}_u)). \tag{6.1}$$

Here, we assume that $\mathbf{y}_u \in \mathbb{Z}^c$ is a one-hot vector indicating the class of training node $u \in \mathcal{V}_{\text{train}}$; for example, in the citation network setting, $\mathbf{y}_u$ would indicate the topic of paper u. We use $\text{softmax}(\mathbf{z}_u, \mathbf{y}_u)$ to denote the predicted probability that the node belongs to the class $\mathbf{y}_u$, computed via the softmax function:

$$\text{softmax}(\mathbf{z}_u, \mathbf{y}_u) = \sum_{i=1}^{c} \mathbf{y}_u[i] \frac{e^{\mathbf{z}_u^\top \mathbf{w}_i}}{\sum_{j=1}^{c} e^{\mathbf{z}_u^\top \mathbf{w}_j}}, \tag{6.2}$$

where $\mathbf{w}_i \in \mathbb{R}^d, i = 1, ..., c$ are trainable parameters. There are other variations of supervised node losses, but training GNNs in a supervised manner based on the loss in Equation (6.1) is one of the most common optimization strategies for GNNs.

Supervised, semi-supervised, transductive, and inductive   Note that—as discussed in Chapter 1—the node classification setting is often referred to both as supervised and semi-supervised. One important factor when applying these terms is whether and how different nodes are used during training the GNN. Generally, we can distinguish between three types of nodes:

1. There is the set of training nodes, $\mathcal{V}_{\text{train}}$. These nodes are included in the GNN message passing operations, and they are also used to compute the loss, e.g., via Equation (6.1).

2. In addition to the training nodes, we can also have *transductive* test nodes, $\mathcal{V}_{\text{trans}}$. These nodes are unlabeled and not used in the loss computation, but these nodes—and their incident edges—are still involved in the GNN message passing operations. In other words, the GNN will generate hidden representations $\mathbf{h}_u^{(k)}$ for the nodes in $u \in \mathcal{V}_{\text{trans}}$ during the GNN message passing operations. However, the final layer embeddings $\mathbf{z}_u$ for these nodes will not be used in the loss function computation.

3. Finally, we will also have *inductive* test nodes, $\mathcal{V}_{\text{ind}}$. These nodes are not used in either the loss computation or the GNN message passing operations during training, meaning that these nodes—and all of their edges—are completely unobserved while the GNN is trained.

The term semi-supervised is applicable in cases where the GNN is tested on transductive test nodes, since in this case the GNN observes the test nodes (but not their labels) during training. The term inductive node classification is used to distinguish the setting where the test nodes—and all their incident edges—are completely unobserved during training. An example of inductive node classification would be training a GNN on one subgraph of a citation network and then testing it on a completely disjoint subgraph.

6.1.2   GNNS FOR GRAPH CLASSIFICATION

Similar to node classification, applications on graph-level classification are popular as benchmark tasks. Historically, kernel methods were popular for graph classification, and—as a result—some of the most popular early benchmarks for graph classification were adapted from the kernel literature, such as tasks involving the classification of enzyme properties based on graph-based representations [Morris et al., 2019]. In these tasks, a softmax classification loss—analogous to Equation (6.1)—is often used, with the key difference that the loss is computed with graph-level embeddings $\mathbf{z}_{\mathcal{G}_i}$ over a set of labeled training graphs $\mathcal{T} = \{\mathcal{G}_1, ..., \mathcal{G}_n\}$. In recent years, GNNs have also witnessed success in regression tasks involving graph data—especially tasks involving the prediction of molecular properties (e.g., solubility) from graph-based representations of molecules. In these instances, it is standard to employ a squared-error loss of the following form:

$$\mathcal{L} = \sum_{\mathcal{G}_i \in \mathcal{T}} \|\mathrm{MLP}(\mathbf{z}_{\mathcal{G}_i}) - y_{\mathcal{G}_i}\|_2^2, \tag{6.3}$$

where MLP is a densely connected neural network with a univariate output and $y_{\mathcal{G}_i} \in \mathbb{R}$ is the target value for training graph $\mathcal{G}_i$.

6.1.3   GNNS FOR RELATION PREDICTION

While classification tasks are by far the most popular application of GNNs, GNNs are also used in in relation prediction tasks, such as recommender systems [Ying et al., 2018a] and knowledge graph completion [Schlichtkrull et al., 2017]. In these applications, the standard practice is to employ the pairwise node embedding loss functions introduced in Chapters 3 and 4. In principle, GNNs can be combined with any of the pairwise loss functions discussed in those chapters, with the output of the GNNs replacing the shallow embeddings.

6.1.4   PRE-TRAINING GNNS

Pre-training techniques have become standard practice in deep learning [Goodfellow et al., 2016]. In the case of GNNs, one might imagine that pre-training a GNN using one of the neighborhood reconstruction losses from Chapter 3 could be a useful strategy to improve performance on a downstream classification task. For example, one could pre-train a GNN to reconstruct missing edges in the graph before fine-tuning on a node classification loss.

Interestingly, however, this approach has achieved little success in the context of GNNs. In fact, Veličković et al. [2019] even find that a *randomly initialized GNN* is equally strong compared to one pre-trained on a neighborhood reconstruction loss. One hypothesis to explain this finding is the fact that the GNN message passing already effectively encodes neighborhood information. Neighboring nodes in the graph will tend to have similar embeddings in a GNN due to the structure of message passing, so enforcing a neighborhood reconstruction loss can simply be redundant.

Despite this negative result regarding pre-training with neighborhood reconstruction losses, there have been positive results using other pre-training strategies. For example, Veličković et al. [2019] propose *Deep Graph Infomax (DGI)*, which involves maximizing the mutual information between node embeddings $\mathbf{z}_u$ and graph embeddings $\mathbf{z}_{\mathcal{G}}$. Formally, this approach optimizes the following loss:

$$\mathcal{L} = -\sum_{u \in \mathcal{V}_{\text{train}}} \mathbb{E}_{\mathcal{G}} \log(D(\mathbf{z}_u, \mathbf{z}_{\mathcal{G}})) + \gamma \mathbb{E}_{\tilde{\mathcal{G}}} \log(1 - D(\tilde{\mathbf{z}}_u, \mathbf{z}_{\mathcal{G}})). \tag{6.4}$$

Here, $\mathbf{z}_u$ denotes the embedding of node u generated from the GNN based on graph $\mathcal{G}$, while $\tilde{\mathbf{z}}_u$ denotes an embedding of node u generated based on a *corrupted* version of graph $\mathcal{G}$, denoted $\tilde{\mathcal{G}}$. We use D to denote a *discriminator* function, which is a neural network trained to predict whether the node embedding came from the real graph $\mathcal{G}$ or the corrupted version $\tilde{\mathcal{G}}$. Usually, the graph is corrupted by modifying either the node features, adjacency matrix, or both in some stochastic manner (e.g., shuffling entries of the feature matrix). The intuition behind this loss is that the GNN model must learn to generate node embeddings that can distinguish between the real graph and its corrupted counterpart. It can be shown that this optimization is closely connected to maximizing the mutual information between the node embeddings $\mathbf{z}_u$ and the graph-level embedding $\mathbf{z}_{\mathcal{G}}$.

The loss function used in DGI (Equation (6.4)) is just one example of a broader class of unsupervised objectives that have witnessed success in the context of GNNs [Hu et al., 2019, Sun et al., 2020]. These unsupervised training strategies generally involve training GNNs to maximize the mutual information between different levels of representations or to distinguish between real and corrupted pairs of embeddings. Conceptually, these pre-training approaches—which are also sometimes used as auxiliary losses during supervised training—bear similarities to the "content masking" pre-training approaches that have ushered in a new state of the art in natural language processing [Devlin et al., 2018]. Nonetheless, the extension and improvement of GNN pre-training approaches is an open and active area of research.

6.2   EFFICIENCY CONCERNS AND NODE SAMPLING

In Chapter 5, we mainly discussed GNNs from the perspective of node-level message passing equations. However, directly implementing a GNN based on these equations can be computationally inefficient. For example, if multiple nodes share neighbors, we might end up doing redundant computation if we implement the message passing operations independently for all nodes in the graph. In this section, we discuss some strategies that can be used implement GNNs in an efficient manner.

6.2.1   GRAPH-LEVEL IMPLEMENTATIONS

In terms of minimizing the number of mathematical operations needed to run message passing, the most effective strategy is to use graph-level implementations of the GNN equations. We discussed these graph-level equations in Section 5.1.3 of the previous chapter, and the key idea is to implement the message passing operations based on sparse matrix multiplications. For example, the graph-level equation for a basic GNN is given by

$$\mathbf{H}^{(k)} = \sigma \left(\mathbf{A}\mathbf{H}^{(k-1)}\mathbf{W}^{(k)}_{\text{neigh}} + \mathbf{H}^{(k-1)}\mathbf{W}^{(k)}_{\text{self}} \right), \tag{6.5}$$

where $\mathbf{H}^{(t)}$ is a matrix containing the layer-k embeddings of all the nodes in the graph. The benefit of using these equations is that there are no redundant computations—i.e., we compute the embedding $\mathbf{h}^{(k)}_u$ for each node u exactly once when running the model. However, the limitation of this approach is that it requires operating on the entire graph and all node features simultaneously, which may not be feasible due to memory limitations. In addition, using the graph-level equations essentially limits one to full-batch (as opposed to mini-batched) gradient descent.

6.2.2   SUBSAMPLING AND MINI-BATCHING

In order to limit the memory footprint of a GNN and facilitate mini-batch training, one can work with a subset of nodes during message passing. Mathematically, we can think of this as running the node-level GNN equations for a subset of the nodes in the graph in each batch. Redundant computations can be avoided through careful engineering to ensure that we only compute the embedding $\mathbf{h}^{(k)}_u$ for each node u in the batch at most once when running the model.

The challenge, however, is that we cannot simply run message passing on a subset of the nodes in a graph without losing information. Every time we remove a node, we also delete its edges (i.e., we modify the adjacency matrix). There is no guarantee that selecting a random subset of nodes will even lead to a connected graph, and selecting a random subset of nodes for each mini-batch can have a severely detrimental impact on model performance.

Hamilton et al. [2017b] propose one strategy to overcome this issue by subsampling node neighborhoods. The basic idea is to first select a set of target nodes for a batch and then to recursively sample the neighbors of these nodes in order to ensure that the connectivity of the graph is maintained. In order to avoid the possibility of sampling too many nodes for a batch, Hamilton et al. [2017b] propose to subsample the neighbors of each node, using a fixed sample size to improve the efficiency of batched tensor operations. Additional subsampling ideas have been proposed in follow-up work [Chen et al., 2018], and these approaches are crucial in making GNNs scalable to massive real-world graphs [Ying et al., 2018a].

6.3   PARAMETER SHARING AND REGULARIZATION

Regularization is a key component of any machine learning model. In the context of GNNs, many of the standard regularization approaches are known to work well, including L2 regularization, dropout [Srivastava et al., 2014], and layer normalization [Ba et al., 2016]. However, there are also regularization strategies that are somewhat specific to the GNN setting.

Parameter Sharing Across Layers

One strategy that is often employed in GNNs with many layers of message passing is parameter sharing. The core idea is to use the same parameters in all the AGGREGATE and UPDATE functions in the GNN. Generally, this approach is most effective in GNNs with more than six layers, and it is often used in conjunction with gated update functions (see Chapter 5) [Li et al., 2015, Selsam et al., 2019].

Edge Dropout

Another GNN-specific strategy is known as *edge dropout*. In this regularization strategy, we randomly remove (or mask) edges in the adjacency matrix during training, with the intuition that this will make the GNN less prone to overfitting and more robust to noise in the adjacency matrix. This approach has been particularly successful in the application of GNNs to knowledge graphs [Schlichtkrull et al., 2017, Teru et al., 2020], and it was an essential technique used in the original graph attention network (GAT) work [Veličković et al., 2018]. Note also that the neighborhood subsampling approaches discussed in Section 6.2.2 lead to this kind of regularization as a side effect, making it a very common strategy in large-scale GNN applications.

CHAPTER 7

Theoretical Motivations

In this chapter, we will visit some of the theoretical underpinnings of GNNs. One of the most intriguing aspects of GNNs is that they were independently developed from distinct theoretical motivations. From one perspective, GNNs were developed based on the theory of graph signal processing, as a generalization of Euclidean convolutions to the non-Euclidean graph domain [Bruna et al., 2014]. At the same time, however, neural message passing approaches—which form the basis of most modern GNNs—were proposed by analogy to message passing algorithms for probabilistic inference in graphical models [Dai et al., 2016]. And lastly, GNNs have been motivated in several works based on their connection to the Weisfeiler–Lehman graph isomorphism test [Hamilton et al., 2017b].

This convergence of three disparate areas into a single algorithm framework is remarkable. That said, each of these three theoretical motivations comes with its own intuitions and history, and the perspective one adopts can have a substantial impact on model development. Indeed, it is no accident that we deferred the description of these theoretical motivations until *after* the introduction of the GNN model itself. In this chapter, our goal is to introduce the key ideas underlying these different theoretical motivations, so that an interested reader is free to explore and combine these intuitions and motivations as they see fit.

7.1 GNNS AND GRAPH CONVOLUTIONS

In terms of research interest and attention, the derivation of GNNs based on connections to graph convolutions is the dominant theoretical paradigm. In this perspective, GNNs arise from the question: How can we generalize the notion of convolutions to general graph-structured data?

7.1.1 CONVOLUTIONS AND THE FOURIER TRANSFORM

In order to generalize the notion of a convolution to graphs, we first must define what we wish to generalize and provide some brief background details. Let f and h be two functions. We can define the general continuous convolution operation $\star$ as

$$(f \star h)(\mathbf{x}) = \int_{\mathbb{R}^d} f(\mathbf{y})h(\mathbf{x} - \mathbf{y})d\mathbf{y}. \tag{7.1}$$

One critical aspect of the convolution operation is that it can be computed by an element-wise product of the *Fourier transforms* of the two functions:

$$(f \star h)(\mathbf{x}) = \mathcal{F}^{-1}\left(\mathcal{F}(f(\mathbf{x})) \circ \mathcal{F}(h(\mathbf{x}))\right), \tag{7.2}$$

where

$$\mathcal{F}(f(\mathbf{x})) = \hat{f}(\mathbf{s}) = \int_{\mathbb{R}^d} f(\mathbf{x}) e^{-2\pi \mathbf{x}^\top \mathbf{s} i} \, d\mathbf{x} \tag{7.3}$$

is the Fourier transform of $f(\mathbf{x})$ and its inverse Fourier transform is defined as

$$\mathcal{F}^{-1}(\hat{f}(\mathbf{s})) = \int_{\mathbb{R}^d} \hat{f}(\mathbf{s}) e^{2\pi \mathbf{x}^\top \mathbf{s} i} \, d\mathbf{s}. \tag{7.4}$$

In the simple case of univariate discrete data over a finite domain $t \in \{0, .., N-1\}$ (i.e., restricting to finite impulse response filters) we can simplify these operations to a discrete circular convolution[1]

$$(f \star_N h)(t) = \sum_{\tau=0}^{N-1} f(\tau) h((t-\tau)_{\mathrm{mod}\, N}) \tag{7.5}$$

and a discrete Fourier transform (DFT)

$$s_k = \frac{1}{\sqrt{N}} \sum_{t=0}^{N-1} f(x_t) e^{-\frac{i2\pi}{N}kt} \tag{7.6}$$

$$= \frac{1}{\sqrt{N}} \sum_{t=0}^{N-1} f(x_t) \left(\cos\left(\frac{2\pi}{N} kt\right) - i \sin\left(\frac{2\pi}{N} kt\right) \right), \tag{7.7}$$

where $s_k \in \{s_0..., s_{N-1}\}$ is the Fourier coefficient corresponding to the sequence $(f(x_0), f(x_1), ..., f(x_{N-1}))$. In Equation (7.5) we use the notation $\star_N$ to emphasize that this is a circular convolution defined over the finite domain $\{0, ..., N-1\}$, but we will often omit this subscript for notational simplicity.

Interpreting the (discrete) Fourier transform   The Fourier transform essentially tells us how to represent our input signal as a weighted sum of (complex) sinusoidal waves. If we assume that both the input data and its Fourier transform are real-valued, we can interpret the sequence $[s_0, s_1, ..., s_{N-1}]$ as the coefficients of a Fourier series. In this view, s_k tells us the amplitude of the complex sinusoidal component $e^{-\frac{i2\pi}{N}k}$, which has frequency $\frac{2\pi k}{N}$ (in radians). Often we will discuss *high-frequency components* that have a large k and vary quickly as well as *low-frequency components* that have $k \ll N$ and vary more slowly. This notion of

---

[1]For simplicity, we limit ourselves to finite support for both f and h and define the boundary condition using a modulus operator and circular convolution.

low- and high-frequency components will also have an analog in the graph domain, where we will consider signals propagating between nodes in the graph.

In terms of signal processing, we can view the discrete convolution $f \star h$ as a *filtering operation* of the series $(f(x_1), f(x_2), ..., f(x_N))$ by a filter h. Generally, we view the series as corresponding to the values of the signal throughout time, and the convolution operator applies some filter (e.g., a band-pass filter) to modulate this time-varying signal.

One critical property of convolutions, which we will rely on below, is the fact that they are *translation (or shift) equivariant*:

$$f(t + a) \star g(t) = f(t) \star g(t + a) = (f \star g)(t + a). \qquad (7.8)$$

This property means that translating a signal and then convolving it by a filter is equivalent to convolving the signal and then translating the result. Note that as a corollary convolutions are also equivariant to the difference operation:

$$\Delta f(t) \star g(t) = f(t) \star \Delta g(t) = \Delta(f \star g)(t), \qquad (7.9)$$

where

$$\Delta f(t) = f(t + 1) - f(t) \qquad (7.10)$$

is the Laplace (i.e., difference) operator on discrete univariate signals.

These notions of filtering and translation equivariance are central to digital signal processing (DSP) and also underly the intuition of CNNs, which utilize a discrete convolution on two-dimensional data. We will not attempt to cover even a small fraction of the fields of digital signal processing, Fourier analysis, and harmonic analysis here, and we point the reader to various textbooks on these subjects [Grafakos, 2004, Katznelson, 2004, Oppenheim et al., 1999, Rabiner and Gold, 1975].

7.1.2   FROM TIME SIGNALS TO GRAPH SIGNALS

In the previous section, we (briefly) introduced the notions of filtering and convolutions with respect to discrete time-varying signals. We now discuss how we can connect discrete time-varying signals with signals on a graph. Suppose we have a discrete time-varying signal $f(t_0), f(t_1), ..., f(t_{N-1})$. One way of viewing this signal is as corresponding to a chain (or cycle) graph (Figure 7.1), where each point in time t is represented as a node and each function value $f(t)$ represents the signal value at that time/node. Taking this view, it is convenient to represent the signal as a vector $\mathbf{f} \in \mathbb{R}^N$, with each dimension corresponding to a different node in the chain graph. In other words, we have that $\mathbf{f}[t] = f(t)$ (as a slight abuse of notation). The edges in the graph thus represent how the signal propagates; i.e., the signal propagates forward in time.[2]

---

[2]Note that we add a connection between the last and first nodes in the chain as a boundary condition to keep the domain finite.

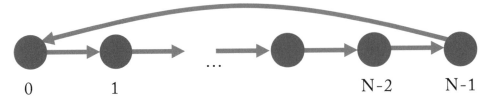

Figure 7.1: Representation of a (cyclic) time-series as a chain graph.

One interesting aspect of viewing a time-varying signal as a chain graph is that we can represent operations, such as time-shifts, using the adjacency and Laplacian matrices of the graph. In particular, the adjacency matrix for this chain graph corresponds to the circulant matrix $\mathbf{A}_c$ with

$$\mathbf{A}_c[i, j] = \begin{cases} 1 & \text{if } j = (i + 1)_{\text{mod } N} \\ 0 & \text{otherwise,} \end{cases} \tag{7.11}$$

and the (unnormalized) Laplacian $\mathbf{L}_c$ for this graph can be defined as

$$\mathbf{L}_c = \mathbf{I} - \mathbf{A}_c. \tag{7.12}$$

We can then represent time shifts as multiplications by the adjacency matrix,

$$(\mathbf{A}_c \mathbf{f})[t] = \mathbf{f}[(t + 1)_{\text{mod } N}], \tag{7.13}$$

and the difference operation by multiplication by the Laplacian,

$$(\mathbf{L}_c \mathbf{f})[t] = \mathbf{f}[(t + 1)_{\text{mod } N}] - \mathbf{f}[t]. \tag{7.14}$$

In this way, we can see that there is a close connection between the adjacency and Laplacian matrices of a graph, and the notions of shifts and differences for a signal. Multiplying a signal by the adjacency matrix propagates signals from node to node, and multiplication by the Laplacian computes the difference between a signal at each node and its immediate neighbors.

Given this graph-based view of transforming signals through matrix multiplication, we can similarly represent convolution by a filter h as matrix multiplication on the vector $\mathbf{f}$:

$$(f \star h)(\mathbf{t}) = \sum_{\tau=0}^{N-1} f(t - \tau)h(\tau) \tag{7.15}$$

$$= \mathbf{Q}_h \mathbf{f}, \tag{7.16}$$

where $\mathbf{Q}_h \in \mathbb{R}^{N \times N}$ is a matrix representation of the convolution operation by filter function h and $\mathbf{f} = [f(t_0), f(t_2), ..., f(t_{N-1})]^\top$ is a vector representation of the function f. Thus, in this view, we consider convolutions that can be represented as a matrix transformation of the signal at each node in the graph.[3] Of course, to have the equality between Equation (7.15) and Equation (7.16), the matrix $\mathbf{Q}_h$ must have some specific properties. In particular, we require that multiplication by this matrix satisfies translation equivariance, which corresponds to commutativity with the circulant adjacency matrix $\mathbf{A}_c$, i.e., we require that

$$\mathbf{A}_c \mathbf{Q}_h = \mathbf{Q}_h \mathbf{A}_c. \tag{7.17}$$

The equivariance to the difference operator is similarly defined as

$$\mathbf{L}_c \mathbf{Q}_h = \mathbf{Q}_h \mathbf{L}_c. \tag{7.18}$$

It can be shown that these requirements are satisfied for a real matrix $\mathbf{Q}_h$ if

$$\mathbf{Q}_h = p_N(\mathbf{A}_c) = \sum_{i=0}^{N-1} \alpha_i \mathbf{A}_c^i, \tag{7.19}$$

i.e., if $\mathbf{Q}_h$ is a polynomial function of the adjacency matrix $\mathbf{A}_c$. In digital signal processing terms, this is equivalent to the idea of representing general filters as polynomial functions of the shift operator [Ortega et al., 2018].[4]

Generalizing to General Graphs

We have now seen how shifts and convolutions on time-varying discrete signals can be represented based on the adjacency matrix and Laplacian matrix of a chain graph. Given this view, we can easily generalize these notions to more general graphs.

In particular, we saw that a time-varying discrete signal corresponds to a chain graph and that the notion of translation/difference equivariance corresponds to a commutativity property with adjacency/Laplacian of this chain graph. Thus, we can generalize these notions beyond the chain graph by considering arbitrary adjacency matrices and Laplacians. While the signal simply propagates forward in time in a chain graph, in an arbitrary graph we might have multiple nodes propagating signals to each other, depending on the structure of the adjacency matrix. Based on this idea, we can define convolutional filters on general graphs as matrices $\mathbf{Q}_h$ that commute with the adjacency matrix or the Laplacian.

More precisely, for an arbitrary graph with adjacency matrix $\mathbf{A}$, we can represent convolutional filters as matrices of the following form:

$$\mathbf{Q}_h = \alpha_0 \mathbf{I} + \alpha_1 \mathbf{A} + \alpha_2 \mathbf{A}^2 + \ldots + \alpha_N \mathbf{A}^N. \tag{7.20}$$

Intuitively, this gives us a *spatial* construction of a convolutional filter on graphs. In particular, if we multiply a node feature vector $\mathbf{x} \in \mathbb{R}^{|V|}$ by such a convolution matrix $\mathbf{Q}_h$, then we get

$$\mathbf{Q}_h \mathbf{x} = \alpha_0 \mathbf{I} \mathbf{x} + \alpha_1 \mathbf{A} \mathbf{x} + \alpha_2 \mathbf{A}^2 \mathbf{x} + \ldots + \alpha_N \mathbf{A}^N \mathbf{x}, \tag{7.21}$$

---

[3]This assumes a real-valued filter h.

[4]Note, however, that there are certain convolutional filters (e.g., complex-valued filters) that cannot be represented in this way.

which means that the convolved signal $\mathbf{Q}_h\mathbf{x}[u]$ at each node $u \in \mathcal{V}$ will correspond to some mixture of the information in the node's N-hop neighborhood, with the $\alpha_0, ..., \alpha_N$ terms controlling the strength of the information coming from different hops.

We can easily generalize this notion of a graph convolution to higher-dimensional node features. If we have a matrix of node features $\mathbf{X} \in \mathbb{R}^{|\mathcal{V}| \times m}$ then we can similarly apply the convolutional filter as

$$\mathbf{Q}_h\mathbf{X} = \alpha_0\mathbf{IX} + \alpha_1\mathbf{AX} + \alpha_2\mathbf{A}^2\mathbf{X} + ... + \alpha_N\mathbf{A}^N\mathbf{X}. \tag{7.22}$$

From a signal processing perspective, we can view the different dimensions of the node features as different "channels."

Graph Convolutions and Message Passing GNNs

Equation (7.22) also reveals the connection between the message passing GNN model we introduced in Chapter 5 and graph convolutions. For example, in the basic GNN approach (see Equation (6.5)) each layer of message passing essentially corresponds to an application of the simple convolutional filter

$$\mathbf{Q}_h = \mathbf{I} + \mathbf{A} \tag{7.23}$$

combined with some learnable weight matrices and a nonlinearity. In general, each layer of message passing GNN architecture aggregates information from a node's local neighborhood and combines this information with the node's current representation (see Equation (5.4)). We can view these message passing layers as a generalization of the simple linear filter in Equation (7.23), where we use more complex nonlinear functions. Moreover, by stacking multiple message passing layers, GNNs are able to implicitly operate on higher-order polynomials of the adjacency matrix.

> **The adjacency matrix, Laplacian, or a normalized variant?** In Equation (7.22) we defined a convolution matrix $\mathbf{Q}_h$ for arbitrary graphs as a polynomial of the adjacency matrix. Defining $\mathbf{Q}_h$ in this way guarantees that our filter commutes with the adjacency matrix, satisfying a generalized notion of *translation equivariance*. However, in general commutativity with the adjacency matrix (i.e., *translation equivariance*) does not necessarily imply commutativity with the Laplacian $\mathbf{L} = \mathbf{D} - \mathbf{A}$ (or any of its normalized variants). In this special case of the chain graph, we were able to define filter matrices $\mathbf{Q}_h$ that simultaneously commute with both $\mathbf{A}$ and $\mathbf{L}$, but for more general graphs we have a choice to make in terms of whether we define convolutions based on the adjacency matrix or some version of the Laplacian. Generally, there is no "right" decision in this case, and there can be empirical trade-offs depending on the choice that is made. Understanding the theoretical underpinnings of these trade-offs is an open area of research [Ortega et al., 2018].
>
> In practice, researchers often use the symmetric normalized Laplacian $\mathbf{L}_{\text{sym}} = \mathbf{D}^{-\frac{1}{2}}\mathbf{L}\mathbf{D}^{-\frac{1}{2}}$ or the symmetric normalized adjacency matrix $\mathbf{A}_{\text{sym}} = \mathbf{D}^{-\frac{1}{2}}\mathbf{A}\mathbf{D}^{-\frac{1}{2}}$ to define

convolutional filters. There are two reasons why these symmetric normalized matrices are desirable. First, both these matrices have bounded spectrums, which gives them desirable numerical stability properties. In addition—and perhaps more importantly—these two matrices are *simultaneously diagonalizable*, which means that they share the same eigenvectors. In fact, one can easily verify that there is a simple relationship between their eigendecompositions, since

$$\mathbf{L}_{\text{sym}} = \mathbf{I} - \mathbf{A}_{\text{sym}}$$
$$\Rightarrow$$
$$\mathbf{L}_{\text{sym}} = \mathbf{U}\mathbf{\Lambda}\mathbf{U}^\top \qquad \mathbf{A}_{\text{sym}} = \mathbf{U}(\mathbf{I} - \mathbf{\Lambda})\mathbf{U}^\top, \qquad (7.24)$$

where $\mathbf{U}$ is the shared set of eigenvectors and $\mathbf{\Lambda}$ is the diagonal matrix containing the Laplacian eigenvalues. This means that defining filters based on one of these matrices implies commutativity with the other, which is a very convenient and desirable property.

7.1.3   SPECTRAL GRAPH CONVOLUTIONS

We have seen how to generalize the notion of a signal and a convolution to the graph domain. We did so by analogy to some important properties of discrete convolutions (e.g., translation equivariance), and this discussion led us to the idea of representing graph convolutions as polynomials of the adjacency matrix (or the Laplacian). However, one key property of convolutions that we ignored in the previous subsection is the relationship between convolutions and the Fourier transform. In this section, we will thus consider the notion of a *spectral convolution* on graphs, where we construct graph convolutions via an extension of the Fourier transform to graphs. We will see that this spectral perspective recovers many of the same results we previously discussed, while also revealing some more general notions of a graph convolution.

The Fourier Transform and the Laplace Operator

To motivate the generalization of the Fourier transform to graphs, we rely on the connection between the Fourier transform and the Laplace (i.e., difference) operator. We previously saw a definition of the Laplace operator Δ in the case of a simple discrete time-varying signal (Equation (7.10)) but this operator can be generalized to apply to arbitrary smooth functions $f : \mathbb{R}^d \to \mathbb{R}$ as

$$\Delta f(\mathbf{x}) = \nabla^2 f(\mathbf{x}) \qquad (7.25)$$
$$= \sum_{i=1}^{n} \frac{\partial^2 f}{\partial x^2}. \qquad (7.26)$$

This operator computes the *divergence* ∇ of the *gradient* $\nabla f(\mathbf{x})$. Intuitively, the Laplace operator tells us the average difference between the function value at a point and function values in the neighboring regions surrounding this point.

In the discrete time setting, the Laplace operator simply corresponds to the difference operator (i.e., the difference between consecutive time points). In the setting of general discrete graphs, this notion corresponds to the Laplacian, since by definition

$$(\mathbf{Lx})[i] = \sum_{j \in \mathcal{V}} \mathbf{A}[i, j](\mathbf{x}[i] - \mathbf{x}[j]), \tag{7.27}$$

which measures the difference between the value of some signal $\mathbf{x}[i]$ at a node i and the signal values of all of its neighbors. In this way, we can view the Laplacian matrix as a discrete analog of the Laplace operator, since it allows us to quantify the difference between the value at a node and the values at that node's neighbors.

Now, an extremely important property of the Laplace operator is that its eigenfunctions correspond to the complex exponentials. That is,

$$-\Delta(e^{2\pi i s t}) = -\frac{\partial^2 (e^{2\pi i s t})^2}{\partial t^2} = (2\pi s)^2 e^{2\pi i s t}, \tag{7.28}$$

so the eigenfucntions of $-\Delta e^{2\pi i s t}$ are the same complex exponentials that make up the modes of the frequency domain in the Fourier transform (i.e., the sinusoidal plane waves), with the corresponding eigenvalue indicating the frequency. In fact, one can even verify that the eigenvectors $\mathbf{u}_1, ..., \mathbf{u}_n$ of the circulant Laplacian $\mathbf{L}_c \in \mathbb{R}^{n \times n}$ for the chain graph are $\mathbf{u}_j = \frac{1}{\sqrt{n}}[1, \omega_j, \omega_j^2, ..., \omega_j^n]$ where $\omega_j = e^{\frac{2\pi j}{n}}$.

The graph Fourier transform

The connection between the eigenfunctions of the Laplace operator and the Fourier transform allows us to generalize the Fourier transform to arbitrary graphs. In particular, we can generalize the notion of a Fourier transform by considering the eigendecomposition of the general graph Laplacian:

$$\mathbf{L} = \mathbf{U}\mathbf{\Lambda}\mathbf{U}^{\top}, \tag{7.29}$$

where we define the eigenvectors $\mathbf{U}$ to be the *graph Fourier modes*, as a graph-based notion of Fourier modes. The matrix $\mathbf{\Lambda}$ is assumed to have the corresponding eigenvalues along the diagonal, and these eigenvalues provide a graph-based notion of different frequency values. In other words, since the eigenfunctions of the general Laplace operator correspond to the Fourier modes—i.e., the complex exponentials in the Fourier series—we define the Fourier modes for a general graph based on the eigenvectors of the graph Laplacian.

Thus, the Fourier transform of signal (or function) $\mathbf{f} \in \mathbb{R}^{|\mathcal{V}|}$ on a graph can be computed as

$$\mathbf{s} = \mathbf{U}^{\top}\mathbf{f} \tag{7.30}$$

and its inverse Fourier transform computed as

$$\mathbf{f} = \mathbf{U}\mathbf{s}. \tag{7.31}$$

Graph convolutions in the spectral domain are defined via point-wise products in the transformed Fourier space. In other words, given the graph Fourier coefficients $\mathbf{U}^\top\mathbf{f}$ of a signal $\mathbf{f}$ as well as the graph Fourier coefficients $\mathbf{U}^\top\mathbf{h}$ of some filter $\mathbf{h}$, we can compute a graph convolution via element-wise products as

$$\mathbf{f} \star_{\mathcal{G}} \mathbf{h} = \mathbf{U}\left(\mathbf{U}^\top\mathbf{f} \circ \mathbf{U}^\top\mathbf{h}\right), \tag{7.32}$$

where $\mathbf{U}$ is the matrix of eigenvectors of the Laplacian $\mathbf{L}$ and where we have used $\star_{\mathcal{G}}$ to denote that this convolution is specific to a graph $\mathcal{G}$.

Based on Equation (7.32), we can represent convolutions in the spectral domain based on the graph Fourier coefficients $\theta_h = \mathbf{U}^\top\mathbf{h} \in \mathbb{R}^{|\mathcal{V}|}$ of the function h. For example, we could learn a *nonparametric* filter by directly optimizing θ_h and defining the convolution as

$$\mathbf{f} \star_{\mathcal{G}} \mathbf{h} = \mathbf{U}\left(\mathbf{U}^\top\mathbf{f} \circ \theta_h\right) \tag{7.33}$$
$$= (\mathbf{U}\mathrm{diag}(\theta_h)\mathbf{U}^\top)\mathbf{f} \tag{7.34}$$

where $\mathrm{diag}(\theta_h)$ is matrix with the values of θ_h on the diagonal. However, a filter defined in this non-parametric way has no real dependency on the structure of the graph and may not satisfy many of the properties that we want from a convolution. For example, such filters can be arbitrarily *non-local*.

To ensure that the spectral filter θ_h corresponds to a meaningful convolution on the graph, a natural solution is to parameterize θ_h based on the eigenvalues of the Laplacian. In particular, we can define the spectral filter as $p_N(\boldsymbol{\Lambda})$, so that it is a degree N polynomial of the eigenvalues of the Laplacian. Defining the spectral convolution in this way ensures our convolution commutes with the Laplacian, since

$$\mathbf{f} \star_{\mathcal{G}} \mathbf{h} = (\mathbf{U}p_N(\boldsymbol{\Lambda})\mathbf{U}^\top)\mathbf{f} \tag{7.35}$$
$$= p_N(\mathbf{L})\mathbf{f}. \tag{7.36}$$

Moreover, this definition ensures a notion of locality. If we use a degree k polynomial, then we ensure that the filtered signal at each node depends on information in its k-hop neighborhood.

Thus, in the end, deriving graph convolutions from the spectral perspective, we can recover the key idea that graph convolutions can be represented as polynomials of the Laplacian (or one of its normalized variants). However, the spectral perspective also reveals more general strategies for defining convolutions on graphs.

Interpreting the Laplacian eigenvectors as frequencies   In the standard Fourier transform we can interpret the Fourier coefficients as corresponding to different frequencies. In the general graph case, we can no longer interpret the graph Fourier transform in this way.

However, we can still make analogies to high-frequency and low-frequency components. In particular, we can recall that the eigenvectors $\mathbf{u}_i, i = 1, ..., |\mathcal{V}|$ of the Laplacian solve the minimization problem:

$$\min_{\mathbf{u}_i \in \mathbb{R}^{|\mathcal{V}|} : \mathbf{u}_i \perp \mathbf{u}_j \forall j < i} \frac{\mathbf{u}_i^\top \mathbf{L} \mathbf{u}_i}{\mathbf{u}_i^\top \mathbf{u}_j} \tag{7.37}$$

by the Rayleigh–Ritz Theorem. And we have that

$$\mathbf{u}_i^\top \mathbf{L} \mathbf{u}_i = \frac{1}{2} \sum_{u,v \in \mathcal{V}} \mathbf{A}[u, v](\mathbf{u}_i[u] - \mathbf{u}_i[v])^2 \tag{7.38}$$

by the properties of the Laplacian discussed in Chapter 1. Together these facts imply that the smallest eigenvector of the Laplacian corresponds to a signal that varies from node to node by the least amount on the graph, the second-smallest eigenvector corresponds to a signal that varies the second smallest amount, and so on. Indeed, we leveraged these properties of the Laplacian eigenvectors in Chapter 1 when we performed spectral clustering. In that case, we showed that the Laplacian eigenvectors can be used to assign nodes to communities so that we minimize the number of edges that go between communities. We can now interpret this result from a signal processing perspective: the Laplacian eigenvectors define signals that vary in a smooth way across the graph, with the smoothest signals indicating the coarse-grained community structure of the graph.

7.1.4   CONVOLUTION-INSPIRED GNNS

The previous subsections generalized the notion of convolutions to graphs. We saw that basic convolutional filters on graphs can be represented as polynomials of the (normalized) adjacency matrix or Laplacian. We saw both spatial and spectral motivations of this fact, and we saw how the spectral perspective can be used to define more general forms of graph convolutions based on the graph Fourier transform. In this section, we will briefly review how different GNN models have been developed and inspired based on these connections.

Purely Convolutional Approaches
Some of the earliest work on GNNs can be directly mapped to the graph convolution definitions of the previous subsections. The key idea in these approaches is that they use either Equation (7.34) or Equation (7.35) to define a convolutional layer, and a full model is defined by stacking and combining multiple convolutional layers with nonlinearities. For example, in early work Bruna et al. [2014] experimented with the nonparametric spectral filter (Equation (7.34)) as well as a parametric spectral filter (Equation (7.35)), where they defined the polynomial $p_N(\mathbf{\Lambda})$ via a cubic spline approach. Following on this work, Defferrard et al. [2016] defined convolutions based on Equation (7.35) and defined $p_N(\mathbf{L})$ using *Chebyshev polynomials*. This

approach benefits from the fact that Chebyshev polynomials have an efficient recursive formulation and have various properties that make them suitable for polynomial approximation [Mason and Handscomb, 2002]. In a related approach, Liao et al. [2019b] learn polynomials of the Laplacian based on the Lanczos algorithm.

There are also approaches that go beyond real-valued polynomials of the Laplacian (or the adjacency matrix). For example, Levie et al. [2018] consider *Cayley polynomials* of the Laplacian and Bianchi et al. [2019] consider *ARMA filters*. Both of these approaches employ more general parametric rational complex functions of the Laplacian (or the adjacency matrix).

Graph Convolutional Networks and Connections to Message Passing

In their seminal work, Kipf and Welling [2016a] built off the notion of graph convolutions to define one of the most popular GNN architectures, commonly known as the graph convolutional network (GCN). The key insight of the GCN approach is that we can build powerful models by stacking very simple graph convolutional layers. A basic GCN layer is defined in Kipf and Welling [2016a] as

$$\mathbf{H}^{(k)} = \sigma\left(\tilde{\mathbf{A}}\mathbf{H}^{(k-1)}\mathbf{W}^{(k)}\right), \tag{7.39}$$

where $\tilde{\mathbf{A}} = (\mathbf{D}+\mathbf{I})^{-\frac{1}{2}}(\mathbf{I}+\mathbf{A})(\mathbf{D}+\mathbf{I})^{-\frac{1}{2}}$ is a normalized variant of the adjacency matrix (with self-loops) and $\mathbf{W}^{(k)}$ is a learnable parameter matrix. This model was initially motivated as a combination of a simple graph convolution (based on the polynomial $\mathbf{I}+\mathbf{A}$), with a learnable weight matrix and a nonlinearity.

As discussed in Chapter 5, we can also interpret the GCN model as a variation of the basic GNN message passing approach. In general, if we consider combining a simple graph convolution defined via the polynomial $\mathbf{I}+\mathbf{A}$ with nonlinearities and trainable weight matrices we recover the basic GNN:

$$\mathbf{H}^{(k)} = \sigma\left(\mathbf{A}\mathbf{H}^{(k-1)}\mathbf{W}^{(k)}_{\text{neigh}} + \mathbf{H}^{(k-1)}\mathbf{W}^{(k)}_{\text{self}}\right). \tag{7.40}$$

In other words, a simple graph convolution based on $\mathbf{I}+\mathbf{A}$ is equivalent to aggregating information from neighbors and combining that with information from the node itself. Thus, we can view the notion of message passing as corresponding to a simple form of graph convolutions combined with additional trainable weights and nonlinearities.

> **Over-smoothing as a low-pass convolutional filter** In Chapter 5, we introduced the problem of *over-smoothing* in GNNs. The intuitive idea in over-smoothing is that after too many rounds of message passing, the embeddings for all nodes begin to look identical and are relatively uninformative. Based on the connection between message-passing GNNs and graph convolutions, we can now understand over-smoothing from the perspective of graph signal processing.

The key intuition is that stacking multiple rounds of message passing in a basic GNN is analogous to applying a low-pass convolutional filter, which produces a smoothed version of the input signal on the graph. In particular, suppose we simplify a basic GNN (Equation (7.40)) to the following update equation:

$$\mathbf{H}^{(k)} = \mathbf{A}_{\text{sym}}\mathbf{H}^{(k-1)}\mathbf{W}^{(k)}. \tag{7.41}$$

Compared to the basic GNN in Equation (7.40), we have simplified the model by removing the nonlinearity and removing addition of the "self" embeddings at each message-passing step. For mathematical simplicity and numerical stability, we will also assume that we are using the symmetric normalized adjacency matrix $\mathbf{A}_{\text{sym}} = \mathbf{D}^{-\frac{1}{2}}\mathbf{A}\mathbf{D}^{-\frac{1}{2}}$ rather than the unnormalized adjacency matrix. This model is similar to the simple GCN approach proposed in Kipf and Welling [2016a] and essentially amounts to taking the average over the neighbor embeddings at each round of message passing.

Now, it is easy to see that after K rounds of message passing based on Equation (7.41), we will end up with a representation that depends on the Kth power of the adjacency matrix:

$$\mathbf{H}^{(K)} = \mathbf{A}_{\text{sym}}^{K}\mathbf{X}\mathbf{W}, \tag{7.42}$$

where $\mathbf{W}$ is some linear operator and $\mathbf{X}$ is the matrix of input node features. To understand the connection between over-smoothing and convolutional filters, we just need to recognize that the multiplication $\mathbf{A}_{\text{sym}}^{K}\mathbf{X}$ of the input node features by a high power of the adjacency matrix can be interpreted as convolutional filter based on the lowest-frequency signals of the graph Laplacian.

For example, suppose we use a large enough value of K such that we have reached the a fixed point of the following recurrence:

$$\mathbf{A}_{\text{sym}}\mathbf{H}^{(K)} = \mathbf{H}^{(K)}. \tag{7.43}$$

One can verify that this fixed point is attainable when using the normalized adjacency matrix, since the dominant eigenvalue of $\mathbf{A}_{\text{sym}}$ is equal to one. We can see that at this fixed point, all the node features will have converged to be completely defined by the dominant eigenvector of $\mathbf{A}_{\text{sym}}$, and more generally, higher powers of $\mathbf{A}_{\text{sym}}$ will emphasize the largest eigenvalues of this matrix. Moreover, we know that the largest eigenvalues of $\mathbf{A}_{\text{sym}}$ correspond to the smallest eigenvalues of its counterpart, the symmetric normalized Laplacian $\mathbf{L}_{\text{sym}}$ (e.g., see Equation (7.24)). Together, these facts imply that multiplying a signal by high powers of $\mathbf{A}_{\text{sym}}$ corresponds to a convolutional filter based on the *lowest* eigenvalues (or frequencies) of $\mathbf{L}_{\text{sym}}$, i.e., it produces a low-pass filter!

Thus, we can see from this simplified model that stacking many rounds of message passing leads to convolutional filters that are low-pass, and—in the worst case—these filters simply converge all the node representations to constant values within connected components on the graph (i.e., the "zero-frequency" of the Laplacian).

Of course, in practice we use more complicated forms of message passing, and this issue is partially alleviated by including each node's previous embedding in the message-passing update step. Nonetheless, it is instructive to understand how stacking "deeper" convolutions on graphs in a naive way can actually lead to simpler, rather than more complex, convolutional filters.

GNNs Without Message Passing

Inspired by connections to graph convolutions, several recent works have also proposed to simplify GNNs by removing the iterative message passing process. In these approaches, the models are generally defined as

$$\mathbf{Z} = \mathrm{MLP}_\theta\left(f(\mathbf{A})\mathrm{MLP}_\phi(\mathbf{X})\right), \tag{7.44}$$

where $f : \mathbb{R}^{|\mathcal{V}|\times|\mathcal{V}|} \to \mathbb{R}^{|\mathcal{V}|\times|\mathcal{V}|}$ is some deterministic function of the adjacency matrix $\mathbf{A}$, MLP denotes a dense neural network, $\mathbf{X} \in \mathbb{R}^{|\mathcal{V}|\times m}$ is the matrix of input node features, and $\mathbf{Z} \in \mathbb{R}^{|\mathcal{V}|\times d}$ is the matrix of learned node representations. For example, in Wu et al. [2019], they define

$$f(\mathbf{A}) = \tilde{\mathbf{A}}^k, \tag{7.45}$$

where $\tilde{\mathbf{A}} = (\mathbf{D} + \mathbf{I})^{-\frac{1}{2}}(\mathbf{A} + \mathbf{I})(\mathbf{D} + \mathbf{I})^{-\frac{1}{2}}$ is the symmetric normalized adjacency matrix (with self-loops added). In a closely related work, Klicpera et al. [2019] defines f by analogy to the personalized PageRank algorithm as[5]

$$f(\mathbf{A}) = \alpha(\mathbf{I} - (1-\alpha)\tilde{\mathbf{A}})^{-1} \tag{7.46}$$

$$= \alpha \sum_{k=0}^{\infty} \left(\mathbf{I} - \alpha\tilde{\mathbf{A}}\right)^k. \tag{7.47}$$

The intuition behind these approaches is that we often do not need to interleave trainable neural networks with graph convolution layers. Instead, we can simply use neural networks to learn feature transformations at the beginning and end of the model and apply a deterministic convolution layer to leverage the graph structure. These simple models are able to outperform more heavily parameterized message passing models (e.g., GATs or GraphSAGE) on many classification benchmarks.

There is also increasing evidence that using the symmetric normalized adjacency matrix with self-loops leads to effective graph convolutions, especially in this simplified setting without message passing. Both Wu et al. [2019] and Klicpera et al. [2019] found that convolutions based on $\tilde{\mathbf{A}}$ achieved the best empirical performance. Wu et al. [2019] also provide theoretical support

---

[5]Note that the equality between Equations (7.46) and (7.47) requires that the dominant eigenvalue of $(\mathbf{I} - \alpha\mathbf{A})$ is bounded above by 1. In practice, Klicpera et al. [2019] use power iteration to approximate the inversion in Equation (7.46).

for these results. They prove that adding self-loops shrinks the spectrum of corresponding graph Laplacian by reducing the magnitude of the dominant eigenvalue. Intuitively, adding self-loops decreases the influence of far-away nodes and makes the filtered signal more dependent on local neighborhoods on the graph.

7.2   GNNS AND PROBABILISTIC GRAPHICAL MODELS

GNNs are well understood and well motivated as extensions of convolutions to graph-structured data. However, there are alternative theoretical motivations for the GNN framework that can provide interesting and novel perspectives. One prominent example is the motivation of GNNs based on connections to variational inference in probabilistic graphical models (PGMs).

In this probabilistic perspective, we view the embeddings $z_u, \forall u \in \mathcal{V}$ for each node as *latent variables* that we are attempting to infer. We assume that we observe the graph structure (i.e., the adjacency matrix, $\mathbf{A}$) and the input node features, $\mathbf{X}$, and our goal is to infer the underlying latent variables (i.e., the embeddings z_v) that can explain this observed data. The message passing operation that underlies GNNs can then be viewed as a neural network analog of certain message passing algorithms that are commonly used for variational inference to infer distributions over latent variables. This connection was first noted by Dai et al. [2016], and much of the proceeding discussions is based closely on their work.

Note that the presentation in this section assumes a substantial background in PGMs, and we recommend Wainwright and Jordan [2008] as a good resource for the interested reader. However, we hope and expect that even a reader without any knowledge of PGMS can glean useful insights from the following discussions.

7.2.1   HILBERT SPACE EMBEDDINGS OF DISTRIBUTIONS

To understand the connection between GNNs and probabilistic inference, we first (briefly) introduce the notion of embedding distributions in Hilbert spaces [Smola et al., 2007]. Let $p(\mathbf{x})$ denote a probability density function defined over the random variable $\mathbf{x} \in \mathbb{R}^m$. Given an arbitrary (and possibly infinite dimensional) feature map $\phi : \mathbb{R}^m \to \mathcal{R}$, we can represent the density $p(\mathbf{x})$ based on its expected value under this feature map:

$$\mu_{\mathbf{x}} = \int_{\mathbb{R}^m} \phi(\mathbf{x}) p(\mathbf{x}) d\mathbf{x}. \tag{7.48}$$

The key idea with Hilbert space embeddings of distributions is that Equation (7.48) will be *injective*, as long as a suitable feature map ϕ is used. This means that $\mu_{\mathbf{x}}$ can serve as a *sufficient statistic* for $p(\mathbf{x})$, and any computations we want to perform on $p(\mathbf{x})$ can be equivalently represented as functions of the embedding $\mu_{\mathbf{x}}$. A well-known example of a feature map that would guarantee this injective property is the feature map induced by the Gaussian radial basis function (RBF) kernel [Smola et al., 2007].

The study of Hilbert space embeddings of distributions is a rich area of statistics. In the context of the connection to GNNs, however, the key takeaway is simply that we can represent distributions $p(\mathbf{x})$ as embeddings μ_x in some feature space. We will use this notion to motivate the GNN message passing algorithm as a way of learning embeddings that represent the distribution over node latents $p(\mathbf{z}_v)$.

7.2.2   GRAPHS AS GRAPHICAL MODELS

Taking a probabilistic view of graph data, we can assume that the graph structure we are given defines the *dependencies* between the different nodes. Of course, we usually interpret graph data in this way. Nodes that are connected in a graph are generally assumed to be related in some way. However, in the probabilistic setting, we view this notion of dependence between nodes in a formal, probabilistic way.

To be precise, we say that a graph $\mathcal{G} = (\mathcal{V}, \mathcal{E})$ defines a Markov random field:

$$p(\{\mathbf{x}_v\}, \{\mathbf{z}_v\}) \propto \prod_{v \in V} \Phi(\mathbf{x}_v, \mathbf{z}_v) \prod_{(u,v) \in \mathcal{E}} \Psi(\mathbf{z}_u, \mathbf{z}_v), \tag{7.49}$$

where Φ and Ψ are non-negative *potential functions*, and where we use $\{\mathbf{x}_v\}$ as a shorthand for the set $\{\mathbf{x}_v, \forall v \in \mathcal{V}\}$. Equation (7.49) says that the distribution $p(\{\mathbf{x}_v\}, \{\mathbf{z}_v\})$ over node features and node embeddings factorizes according to the graph structure. Intuitively, $\Phi(\mathbf{x}_v, \mathbf{z}_v)$ indicates the likelihood of a node feature vector $\mathbf{x}_v$ given its latent node embedding $\mathbf{z}_v$, while Ψ controls the dependency between connected nodes. We thus assume that node features are determined by their latent embeddings, and we assume that the latent embeddings for connected nodes are dependent on each other (e.g., connected nodes might have similar embeddings).

In the standard probabilistic modeling setting, Φ and Ψ are usually defined as parametric functions based on domain knowledge, and, most often, these functions are assumed to come from the exponential family to ensure tractability [Wainwright and Jordan, 2008]. In our presentation, however, we are agnostic to the exact form of Φ and Ψ, and we will seek to *implicitly* learn these functions by leveraging the Hilbert space embedding idea discussed in the previous section.

7.2.3   EMBEDDING MEAN-FIELD INFERENCE

Given the Markov random field defined by Equation (7.49), our goal is to infer the distribution of latent embeddings $p(\mathbf{z}_v)$ for all the nodes $v \in V$, while also implicitly learning the potential functions Φ and Ψ. In more intuitive terms, our goal is to infer latent representations for all the nodes in the graph that can explain the dependencies between the observed node features.

In order to do so, a key step is computing the posterior $p(\{\mathbf{z}_v\}|\{\mathbf{x}_v\})$, i.e., computing the likelihood of a particular set of latent embeddings given the observed features. In general, computing this posterior is computationally intractable—even if Φ and Ψ are known and well-defined—so we must resort to approximate methods.

One popular approach—which we will leverage here—is to employ *mean-field variational inference*, where we approximate the posterior using some functions q_v based on the assumption:

$$p(\{\mathbf{z}_v\}|\{\mathbf{x}_v\}) \approx q(\{\mathbf{z}_v\}) = \prod_{v \in \mathcal{V}} q_v(\mathbf{z}_v), \qquad (7.50)$$

where each q_v is a valid density. The key intuition in mean-field inference is that we assume that the posterior distribution over the latent variables factorizes into $\mathcal{V}$ independent distributions, one per node.

To obtain approximating q_v functions that are optimal in the mean-field approximation, the standard approach is to minimize the Kullback–Leibler (KL) divergence between the approximate posterior and the true posterior:

$$\mathrm{KL}(q(\{\mathbf{z}_v\})|\{p(\{\mathbf{z}_v\}|\{\mathbf{x}_v\}) = \int_{(\mathbb{R}^d)^{\mathcal{V}}} \prod_{v \in \mathcal{V}} q(\{\mathbf{z}_v\}) \log \left(\frac{\prod_{v \in \mathcal{V}} q(\{\mathbf{z}_v\})}{p(\{\mathbf{z}_v\}|\{\mathbf{x}_v\})} \right) \prod_{v \in \mathcal{V}} d\mathbf{z}_v. \qquad (7.51)$$

The KL divergence is one canonical way of measuring the distance between probability distributions, so finding q_v functions that minimize Equation (7.51) gives an approximate posterior that is as close as possible to the true posterior under the mean-field assumption. Of course, directly minimizing Equation (7.51) is impossible, since evaluating the KL divergence requires knowledge of the true posterior.

Luckily, however, techniques from variational inference can be used to show that $q_v(\mathbf{z}_v)$ that minimize the KL must satisfy the following fixed point equations:

$$\log(q(\mathbf{z}_v)) = c_v + \log(\Phi(\mathbf{x}_v, \mathbf{z}_v)) + \sum_{u \in \mathcal{N}(v)} \int_{\mathbb{R}^d} q_u(\mathbf{z}_u) \log \left(\Psi(\mathbf{z}_u, \mathbf{z}_v) \right) d\mathbf{z}_u, \qquad (7.52)$$

where c_v is a constant that does not depend on $q_v(\mathbf{z}_v)$ or $\mathbf{z}_v$. In practice, we can approximate this fixed point solution by initializing some initial guesses $q_v^{(t)}$ to valid probability distributions and iteratively computing

$$\log \left(q_v^{(t)}(\mathbf{z}_v) \right) = c_v + \log(\Phi(\mathbf{x}_v, \mathbf{z}_v)) + \sum_{u \in \mathcal{N}(v)} \int_{\mathbb{R}^d} q_u^{(t-1)}(\mathbf{z}_u) \log \left(\Psi(\mathbf{z}_u, \mathbf{z}_v) \right) d\mathbf{z}_u. \qquad (7.53)$$

The justification behind Equation (7.52) is beyond the scope of this book. For the purposes of this book, however, the essential ideas are the following.

1. We can approximate the true posterior $p(\{\mathbf{z}_v\}|\{\mathbf{x}_v\})$ over the latent embeddings using the mean-field assumption, where we assume that the posterior factorizes into $|\mathcal{V}|$ independent distributions $p(\{\mathbf{z}_v\}|\{\mathbf{x}_v\}) \approx \prod_{v \in \mathcal{V}} q_v(\mathbf{z}_v)$.

2. The optimal approximation under the mean-field assumption is given by the fixed point in Equation (7.52), where the approximate posterior $q_v(\mathbf{z}_v)$ for each latent node embedding is a function of (i) the node's feature $\mathbf{z}_x$ and (ii) the marginal distributions $q_u(\mathbf{z})_u, \forall u \in \mathcal{N}(v)$ of the node's neighbors' embeddings.

At this point the connection to GNNs begins to emerge. In particular, if we examine the fixed point iteration in Equation (7.53), we see that the updated marginal distribution $q_v^{(t)}(\mathbf{z}_v)$ is a function of the node features $\mathbf{x}_v$ (through the potential function Φ) as well as function of the set of neighbor marginals $\{q_u^{(t-1)}(\mathbf{z}_u), \forall u \in \mathcal{N}(v)\}$ from the previous iteration (through the potential function Ψ). This form of message passing is highly analogous to the message passing in GNNs! At each step, we are updating the values at each node based on the set of values in the node's neighborhood. The key distinction is that the mean-field message passing equations operate over *distributions* rather than *embeddings*, which are used in the standard GNN message passing.

We can make the connection between GNNs and mean-field inference even tighter by leveraging the Hilbert space embeddings that we introduced in Section 7.2.1. Suppose we have some injective feature map ϕ and can represent all the marginals $q_v(\mathbf{z}_v)$ as embeddings

$$\mu_v = \int_{\mathbb{R}^d} q_v(\mathbf{z}_v)\phi(\mathbf{z}_v)d\mathbf{z}_v \in \mathbb{R}^d. \tag{7.54}$$

With these representations, we can re-write the fixed point iteration in Equation (7.52) as

$$\mu_v^{(t)} = \mathbf{c} + f(\mu_v^{(t-1)}, \mathbf{x}_v, \{\mu_u, \forall u \in \mathcal{N}(v)\} \tag{7.55}$$

where f is a vector-valued function. Notice that f *aggregates* information from the set of neighbor embeddings (i.e., $\{\mu_u, \forall u \in \mathcal{N}(v)\}$ and *updates* the node's current representation (i.e., $\mu_v^{(t-1)}$) using this aggregated data. In this way, we can see that embedded mean-field inference exactly corresponds to a form of neural message passing over a graph!

Now, in the usual probabilistic modeling scenario, we would define the potential functions Φ and Ψ, as well as the feature map ϕ, using some domain knowledge. And given some Φ, Ψ, and ψ, we could then try to analytically derive the f function in Equation (7.55) that would allow us to work with an embedded version of mean field inference. However, as an alternative, we can simply try to learn embeddings μ_v in and end-to-end fashion using some supervised signals, and we can define f to be an arbitrary neural network. In other words, rather than specifying a concrete probabilistic model, we can simply learn embeddings μ_v that *could* correspond to *some* probabilistic model. Based on this idea, Dai et al. [2016] define f in an analogous manner to a basic GNN as

$$\mu_v^{(t)} = \sigma\left(\mathbf{W}_{\text{self}}^{(t)}\mathbf{x}_v + \mathbf{W}_{\text{neigh}}^{(t)} \sum_{u \in \mathcal{N}(v)} \mu_u^{(t-1)}\right). \tag{7.56}$$

Thus, at each iteration, the updated Hilbert space embedding for node v is a function of its neighbors' embeddings as well as its feature inputs. And, as with a basic GNN, the parameters $\mathbf{W}_{\text{self}}^{(t)}$ and $\mathbf{W}_{\text{neigh}}^{(t)}$ of the update process can be trained via gradient descent on any arbitrary task loss.

7.2.4   GNNS AND PGMS MORE GENERALLY

In the previous subsection, we gave a brief introduction to how a basic GNN model can be derived as an embedded form of mean field inference—a connection first outlined by Dai et al. [2016]. There are, however, further ways to connect PGMs and GNNs. For example, different variants of message passing can be derived based on different approximate inference algorithms (e.g., Bethe approximations as discussed in Dai et al. [2016]), and there are also several works which explore how GNNs can be integrated more generally into PGM models [Qu et al., 2019, Zhang et al., 2020]. In general, the connections between GNNs and more traditional statistical relational learning is a rich area with vast potential for new developments.

7.3   GNNS AND GRAPH ISOMORPHISM

We have now seen how GNNs can be motivated based on connections to graph signal processing and probabilistic graphical models. In this section, we will turn our attention to our third and final theoretical perspective on GNNs: the motivation of GNNs based on connections to graph isomorphism testing.

As with the previous sections, here we will again see how the basic GNN can be derived as a neural network variation of an existing algorithm—in this case the Weisfieler-Lehman (WL) isomorphism algorithm. However, in addition to motivating the GNN approach, connections to isomorphism testing will also provide us with tools to analyze the power of GNNs in a formal way.

7.3.1   GRAPH ISOMORPHISM

Testing for graph isomorphism is one of the most fundamental and well-studied tasks in graph theory. Given a pair of graphs $\mathcal{G}_1$ and $\mathcal{G}_2$, the goal of graph isomorphism testing is to declare whether or not these two graphs are *isomorphic*. In an intuitive sense, two graphs being isomorphic means that they are essentially identical. Isomorphic graphs represent the exact same graph structure, but they might differ only in the ordering of the nodes in their corresponding adjacency matrices. Formally, if we have two graphs with adjacency matrices $\mathbf{A}_1$ and $\mathbf{A}_2$, as well as node features $\mathbf{X}_1$ and $\mathbf{X}_2$, we say that two graphs are isomorphic if and only if there exists a permutation matrix $\mathbf{P}$ such that

$$\mathbf{P}\mathbf{A}_1\mathbf{P}^\top = \mathbf{A}_2 \text{ and } \mathbf{P}\mathbf{X}_1 = \mathbf{X}_2. \tag{7.57}$$

It is important to note that isomorphic graphs are really are identical in terms of their underlying structure. The ordering of the nodes in the adjacency matrix is an arbitrary decision we must make when we represent a graph using algebraic objects (e.g., matrices), but this ordering has no bearing on the structure of the underlying graph itself.

Despite its simple definition, testing for graph isomorphism is a fundamentally hard problem. For instance, a naive approach to test for isomorphism would involve the following optimization problem:

$$\min_{\mathbf{P} \in \mathcal{P}} \| \mathbf{P} \mathbf{A}_1 \mathbf{P}^\top - \mathbf{A}_2 \| + \| \mathbf{P} \mathbf{X}_1 - \mathbf{X}_2 \| \stackrel{?}{=} 0. \tag{7.58}$$

This optimization requires searching over the full set of permutation matrices $\mathcal{P}$ to evaluate whether or not there exists a single permutation matrix $\mathbf{P}$ that leads to an equivalence between the two graphs. The computational complexity of this naive approach is immense at $O(|V|!)$, and in fact, no polynomial time algorithm is known to correctly test isomorphism for general graphs.

Graph isomorphism testing is formally referred to as NP-indeterminate (NPI). It is known to not be NP-complete, but no general polynomial time algorithms are known for the problem. (Integer factorization is another well-known problem that is suspected to belong to the NPI class.) There are, however, many practical algorithms for graph isomorphism testing that work on broad classes of graphs, including the WL algorithm that we introduced briefly in Chapter 1.

7.3.2   GRAPH ISOMORPHISM AND REPRESENTATIONAL CAPACITY

The theory of graph isomorphism testing is particularly useful for graph representation learning. It gives us a way to quantify the *representational power* of different learning approaches. If we have an algorithm—for example, a GNN—that can generate representations $\mathbf{z}_\mathcal{G} \in \mathbb{R}^d$ for graphs, then we can quantify the power of this learning algorithm by asking how useful these representations would be for testing graph isomorphism. In particular, given learned representations $\mathbf{z}_{\mathcal{G}_1}$ and $\mathbf{z}_{\mathcal{G}_2}$ for two graphs, a "perfect" learning algorithm would have that

$$\mathbf{z}_{\mathcal{G}_1} = \mathbf{z}_{\mathcal{G}_2} \text{ if and only if } \mathcal{G}_1 \text{ is isomorphic to } \mathcal{G}_2. \tag{7.59}$$

A perfect learning algorithm would generate identical embeddings for two graphs if and only if those two graphs were actually isomorphic.

Of course, in practice, no representation learning algorithm is going to be "perfect" (unless P=NP). Nonetheless, quantifying the power of a representation learning algorithm by connecting it to graph isomorphism testing is very useful. Despite the fact that graph isomorphism testing is not solvable in general, we do know several powerful and well-understood approaches for approximate isomorphism testing, and we can gain insight into the power of GNNs by comparing them to these approaches.

7.3.3   THE WEISFIELER–LEHMAN ALGORITHM

The most natural way to connect GNNs to graph isomorphism testing is based on connections to the family of Weisfeiler–Lehman (WL) algorithms. In Chapter 1, we discussed the WL algorithm in the context of graph kernels. However, the WL approach is more broadly known as one of the most successful and well-understood frameworks for approximate isomorphism testing. The simplest version of the WL algorithm—commonly known as the 1-WL—consists of the following steps.

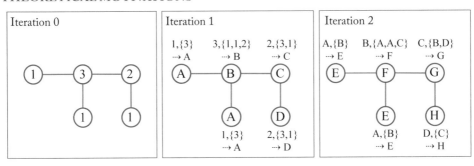

Figure 7.2: Example of the WL iterative labeling procedure on one graph.

1. Given two graphs $\mathcal{G}_1$ and $\mathcal{G}_2$ we assign an initial label $l_{\mathcal{G}_i}^{(0)}(v)$ to each node in each graph. In most graphs, this label is simply the node degree, i.e., $l^{(0)}(v) = d_v \; \forall v \in V$, but if we have discrete features (i.e., one hot features $\mathbf{x}_v$) associated with the nodes, then we can use these features to define the initial labels.

2. Next, we iteratively assign a new label to each node in each graph by hashing the multi-set of the current labels within the node's neighborhood, as well as the node's current label:

$$l_{\mathcal{G}_i}^{(i)}(v) = \text{HASH}\left(l_{\mathcal{G}_i}^{(i-1)}(v), \left\{\!\!\left\{ l_{\mathcal{G}_i}^{(i-1)}(u) \; \forall u \in \mathcal{N}(v) \right\}\!\!\right\}\right), \qquad (7.60)$$

where the double-braces are used to denote a multi-set and the HASH function maps each unique multi-set to a unique new label.

3. We repeat Step 2 until the labels for all nodes in both graphs converge, i.e., until we reach an iteration K where $l_{\mathcal{G}_j}^{(K)}(v) = l_{\mathcal{G}_i}^{(K-1)}(v), \forall v \in V_j, j = 1, 2$.

4. Finally, we construct multi-sets

$$L_{\mathcal{G}_j} = \left\{\!\!\left\{ l_{\mathcal{G}_j}^{(i)}(v), \forall v \in \mathcal{V}_j, i = 0, ..., K-1 \right\}\!\!\right\}$$

summarizing all the node labels in each graph, and we declare $\mathcal{G}_1$ and $\mathcal{G}_2$ to be isomorphic if and only if the multi-sets for both graphs are identical, i.e., if and only if $L_{\mathcal{G}_1} = L_{\mathcal{G}_2}$.

Figure 7.2 illustrates an example of the WL labeling process on one graph. At each iteration, every node collects the multi-set of labels in its local neighborhood, and updates its own label based on this multi-set. After K iterations of this labeling process, every node has a label that summarizes the structure of its K-hop neighborhood, and the collection of these labels can be used to characterize the structure of an entire graph or subgraph.

The WL algorithm is known to converge in at most $|\mathcal{V}|$ iterations and is known to known to successfully test isomorphism for a broad class of graphs [Babai and Kucera, 1979]. There are, however, well-known cases where the test fails, such as the simple example illustrated in Figure 7.3.

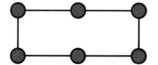 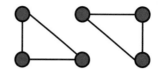

Figure 7.3: Example of two graphs that cannot be distinguished by the basic WL algorithm.

7.3.4   GNNS AND THE WL ALGORITHM

There are clear analogies between the WL algorithm and the neural message passing GNN approach. In both approaches, we iteratively aggregate information from local node neighborhoods and use this aggregated information to update the representation of each node. The key distinction between the two approaches is that the WL algorithm aggregates and updates discrete labels (using a hash function) while GNN models aggregate and update node embeddings using neural networks. In fact, GNNs have been motivated and derived as a continuous and differentiable analog of the WL algorithm.

The relationship between GNNs and the WL algorithm (described in Section 7.3.3) can be formalized in the following theorem:

Theorem 7.1 [**Morris et al., 2019, Xu et al., 2019**].   *Define a message-passing GNN (MP-GNN) to be any GNN that consists of K message-passing layers of the following form:*

$$\mathbf{h}_u^{(k+1)} = \text{UPDATE}^{(k)} \left(\mathbf{h}_u^{(k)}, \text{AGGREGATE}^{(k)}(\{\mathbf{h}_v^{(k)}, \forall v \in \mathcal{N}(u)\}) \right), \tag{7.61}$$

where AGGREGATE *is a differentiable permutation invariant function and* UPDATE *is a differentiable function. Further, suppose that we have only discrete feature inputs at the initial layer, i.e.,* $\mathbf{h}_u^{(0)} = \mathbf{x}_u \in \mathbb{Z}^d, \forall u \in \mathcal{V}$. *Then we have that* $\mathbf{h}_u^{(K)} \neq \mathbf{h}_u^{(K)}$ *only if the nodes u and v have different labels after K iterations of the WL algorithm.*

In intuitive terms, Theorem 7.3.4 states that GNNs are *no more powerful than the WL algorithm* when we have discrete information as node features. If the WL algorithm assigns the same label to two nodes, then any message-passing GNN will also assign the same embedding to these two nodes. This result on node labeling also extends to isomorphism testing. If the WL test cannot distinguish between two graphs, then a MP-GNN is also incapable of distinguishing between these two graphs. We can also show a more positive result in the other direction:

Theorem 7.2 [**Morris et al., 2019, Xu et al., 2019**].   *There exists a MP-GNN such that* $\mathbf{h}_u^{(K)} = \mathbf{h}_v^{(K)}$ *if and only if the two nodes u and v have the same label after K iterations of the WL algorithm.*

This theorem states that there exist message-passing GNNs that are as powerful as the WL test.

Which MP-GNNs are most powerful?   The two theorems above state that message-passing GNNs are at most as powerful as the WL algorithm and that there exist message-passing GNNs that are as powerful the WL algorithm. So which GNNs actually obtain this theoretical upper bound? Interestingly, the basic GNN that we introduced at the beginning of Chapter 5 is sufficient to satisfy this theory. In particular, if we define the message passing updates as follows:

$$\mathbf{h}_u^{(k)} = \sigma \left(\mathbf{W}_{\text{self}}^{(k)} \mathbf{h}_u^{(k-1)} + \mathbf{W}_{\text{neigh}}^{(k)} \sum_{v \in \mathcal{N}(u)} \mathbf{h}_v^{(k-1)} + \mathbf{b}^{(k)} \right), \qquad (7.62)$$

then this GNN is sufficient to match the power of the WL algorithm [Morris et al., 2019].

However, most of the other GNN models discussed in Chapter 5 are *not* as powerful as the WL algorithm. Formally, to be as powerful as the WL algorithm, the AGGREGATE and UPDATE functions need to be *injective* [Xu et al., 2019]. This means that the AGGREGATE and UPDATE operators need to be map every unique input to a unique output value, which is not the case for many of the models we discussed. For example, AGGREGATE functions that use a (weighted) average of the neighbor embeddings are not injective; if all the neighbors have the same embedding then a (weighted) average will not be able to distinguish between input sets of different sizes.

Xu et al. [2019] provide a detailed discussion of the relative power of various GNN architectures. They also define a "minimal" GNN model, which has few parameters but is still as powerful as the WL algorithm. They term this model the *Graph Isomorphism Network (GIN)*, and it is defined by the following update:

$$\mathbf{h}_u^{(k)} = \text{MLP}^{(k)} \left((1 + \epsilon^{(k)}) \mathbf{h}_u^{(k-1)} + \sum_{v \in \mathcal{N}(u)} \mathbf{h}_v^{(k-1)} \right), \qquad (7.63)$$

where $\epsilon^{(k)}$ is a trainable parameter.

7.3.5   BEYOND THE WL ALGORITHM

The previous subsection highlighted an important negative result regarding message-passing GNNs (MP-GNNs): these models are no more powerful than the WL algorithm. However, despite this negative result, investigating how we can make GNNs that are provably *more powerful* than the WL algorithm is an active area of research.

Relational Pooling

One way to motivate provably more powerful GNNs is by considering the failure cases of the WL algorithm. For example, we can see in Figure 7.3 that the WL algorithm—and thus all MP-GNNs—cannot distinguish between a connected six-cycle and a set of two triangles. From the perspective of message passing, this limitation stems from the fact that AGGREGATE and UPDATE operations are unable to detect when two nodes share a neighbor. In the example in Figure 7.3, each node can infer from the message passing operations that they have two degree-2 neighbors, but this information is not sufficient to detect whether a node's neighbors are connected to one another. This limitation is not simply a corner case illustrated in Figure 7.3. Message passing approaches generally fail to identify closed triangles in a graph, which is a critical limitation.

To address this limitation, Murphy et al. [2019] consider augmenting MP-GNNs with *unique node ID features*. If we use MP-GNN($\mathbf{A}, \mathbf{X}$) to denote an arbitrary MP-GNN on input adjacency matrix $\mathbf{A}$ and node features $\mathbf{X}$, then adding node IDs is equivalent to modifying the MP-GNN to the following:

$$\text{MP-GNN}(\mathbf{A}, \mathbf{X} \oplus \mathbf{I}), \tag{7.64}$$

where $\mathbf{I}$ is the $d \times d$-dimensional identity matrix and $\oplus$ denotes column-wise matrix concatenation. In other words, we simple add a unique, one-hot indicator feature for each node. In the case of Figure 7.3, adding unique node IDs would allow a MP-GNN to identify when two nodes share a neighbor, which would make the two graphs distinguishable.

Unfortunately, however, this idea of adding node IDs does not solve the problem. In fact, by adding unique node IDs we have actually introduced a new and equally problematic issue: the MP-GNN is no longer permutation equivariant. For a standard MP-GNN we have that

$$\mathbf{P}(\text{MP-GNN}(\mathbf{A}, \mathbf{X})) = \text{MP-GNN}(\mathbf{P}\mathbf{A}\mathbf{P}^\top, \mathbf{P}\mathbf{X}), \tag{7.65}$$

where $\mathbf{P} \in \mathcal{P}$ is an arbitrary permutation matrix. This means that standard MP-GNNs are permutation equivariant. If we permute the adjacency matrix and node features, then the resulting node embeddings are simply permuted in an equivalent way. However, MP-GNNs with node IDs are not permutation invariant since, in general,

$$\mathbf{P}(\text{MP-GNN}(\mathbf{A}, \mathbf{X} \oplus \mathbf{I})) \neq \text{MP-GNN}(\mathbf{P}\mathbf{A}\mathbf{P}^\top, (\mathbf{P}\mathbf{X}) \oplus \mathbf{I}). \tag{7.66}$$

The key issue is that assigning a unique ID to each node fixes a particular node ordering for the graph, which breaks the permutation equivariance.

To alleviate this issue, Murphy et al. [2019] propose the *Relational Pooling (RP)* approach, which involves marginalizing over all possible node permutations. Given any MP-GNN the RP extension of this GNN is given by

$$\text{RP-GNN}(\mathbf{A}, \mathbf{X}) = \sum_{\mathbf{P} \in \mathcal{P}} \text{MP-GNN}(\mathbf{P}\mathbf{A}\mathbf{P}^\top, (\mathbf{P}\mathbf{X}) \oplus \mathbf{I}). \tag{7.67}$$

Summing over all possible permutation matrices $\mathbf{P} \in \mathcal{P}$ recovers the permutation invariance, and we retain the extra representational power of adding unique node IDs. In fact, Murphy et al. [2019] prove that the RP extension of a MP-GNN can distinguish graphs that are indistinguishable by the WL algorithm.

The limitation of the RP approach is in its computational complexity. Naively evaluating Equation (7.67) has a time complexity of $O(|\mathcal{V}|!)$, which is infeasible in practice. Despite this limitation, however, Murphy et al. [2019] show that the RP approach can achieve strong results using various approximations to decrease the computation cost (e.g., sampling a subset of permutations).

The k-WL Test and k-GNNs

The RP approach discussed above can produce GNN models that are provably more powerful than the WL algorithm. However, the RP approach has two key limitations:

1. The full algorithm is computationally intractable.

2. We know that RP-GNNs are more powerful than the WL test, but we have no way to characterize *how much more* powerful they are.

To address these limitations, several approaches have considered improving GNNs by adapting generalizations of the WL algorithm.

The WL algorithm we introduced in Section 7.3.3 is in fact just the simplest of what is known as the *family of k-WL algorithms*. In fact, the WL algorithm we introduced previously is often referred to as the *1-WL algorithm*, and it can be generalized to the *k-WL algorithm* for $k > 1$. The key idea behind the k-WL algorithms is that we label subgraphs of size k rather than individual nodes. The k-WL algorithm generates representation of a graph $\mathcal{G}$ through the following steps:

1. Let $s = (u_1, u_2, ..., u_k) \in \mathcal{V}^k$ be a tuple defining a subgraph of size k, where $u_1 \neq u_2 \neq ... \neq u_k$. Define the initial label $l_{\mathcal{G}}^{(0)}(s)$ for each subgraph by the isomorphism class of this subgraph (i.e., two subgraphs get the same label if and only if they are isomorphic).

2. Next, we iteratively assign a new label to each subgraph by hashing the multi-set of the current labels within this subgraph's neighborhood:

$$l_{\mathcal{G}}^{(i)}(s) = \text{HASH}\left(\{\{l_{\mathcal{G}}^{(i-1)}(s'), \forall s' \in \mathcal{N}_j(s), j = 1, ..., k\}\}, l_{\mathcal{G}}^{(i-1)}(s)\right),$$

where the jth subgraph neighborhood is defined as

$$\mathcal{N}_j(s) = \{\{(u_1, ..., u_{j-1}, v, u_{j+1}, ..., u_k), \forall v \in \mathcal{V}\}\}. \tag{7.68}$$

3. We repeat Step 2 until the labels for all subgraphs converge, i.e., until we reach an iteration K where $l_{\mathcal{G}}^{(K)}(s) = l_{\mathcal{G}}^{(K-1)}(s)$ for every k-tuple of nodes $s \in \mathcal{V}^k$.

4. Finally, we construct a multi-set

$$L_{\mathcal{G}} = \{\{l_{\mathcal{G}}^{(i)}(v), \forall s \in \mathcal{V}^k, i = 0, ..., K - 1\}\}$$

summarizing all the subgraph labels in the graph.

As with the 1-WL algorithm, the summary $L_{\mathcal{G}}$ multi-set generated by the k-WL algorithm can be used to test graph isomorphism by comparing the multi-sets for two graphs. There are also graph kernel methods based on the k-WL test [Morris et al., 2019], which are analogous to the WL-kernel introduced that was in Chapter 1.

An important fact about the k-WL algorithm is that it introduces a hierarchy of representational capacity. For any $k \geq 2$ we have that the $(k + 1)$-WL test is strictly more powerful than the k-WL test.[6] Thus, a natural question to ask is whether we can design GNNs that are as powerful as the k-WL test for $k > 2$, and, of course, a natural design principle would be to design GNNs by analogy to the k-WL algorithm.

Morris et al. [2019] attempt exactly this: they develop a k-GNN approach that is a differentiable and continuous analog of the k-WL algorithm. k-GNNs learn embeddings associated with subgraphs—rather than nodes—and the message passing occurs according to subgraph neighborhoods (e.g., as defined in Equation (7.68)). Morris et al. [2019] prove that k-GNNs can be as expressive as the k-WL algorithm. However, there are also serious computational concerns for both the k-WL test and k-GNNs, as the time complexity of the message passing explodes combinatorially as k increases. These computational concerns necessitate various approximations to make k-GNNs tractable in practice [Morris et al., 2019].

Invariant and Equivariant k-Order GNNs
Another line of work that is motivated by the idea of building GNNs that are as powerful as the k-WL test are the invariant and equivariant GNNs proposed by Maron et al. [2019]. A crucial aspect of message-passing GNNs (MP-GNNs; as defined in Theorem 7.3.4) is that they are equivariant to node permutations, meaning that

$$\mathbf{P}(\text{MP-GNN}(\mathbf{A}, \mathbf{X})) = \text{MP-GNN}(\mathbf{P}\mathbf{A}\mathbf{P}^\top, \mathbf{P}\mathbf{X}) \tag{7.69}$$

for any permutation matrix $\mathbf{P} \in \mathcal{P}$. This equality says that that permuting the input to an MP-GNN simply results in the matrix of output node embeddings being permuted in an analogous way.

In addition to this notion of equivariance, we can also define a similar notion of permutation *invariance* for MP-GNNs at the graph level. In particular, MP-GNNs can be extended with a POOL : $\mathbb{R}^{|\mathcal{V}| \times d} \to \mathbb{R}$ function (see Chapter 5), which maps the matrix of learned node

---

[6]However, note that running the k-WL requires solving graph isomorphism for graphs of size k, since Step 1 in the k-WL algorithm requires labeling graphs according to their isomorphism type. Thus, running the k-WL for $k > 3$ is generally computationally intractable.

embeddings $\mathbf{Z} \in \mathbb{R}^{|\mathcal{V}| \times d}$ to an embedding $\mathbf{z}_\mathcal{G} \in \mathbb{R}^d$ of the entire graph. In this graph-level setting we have that MP-GNNs are permutation invariant, i.e.,

$$\text{POOL} \left(\text{MP-GNN}(\mathbf{P}\mathbf{A}\mathbf{P}^\top, \mathbf{P}\mathbf{X}) \right) = \text{POOL} \left(\text{MP-GNN}(\mathbf{A}, \mathbf{X}) \right), \qquad (7.70)$$

meaning that the pooled graph-level embedding does not change when different node orderings are used.

Based on this idea of invariance and equivariance, Maron et al. [2019] propose a general form of GNN-like models based on permutation equivariant/invariant tensor operations. Suppose we have an order-$(k + 1)$ tensor $\mathcal{X} \in \mathbb{R}^{|\mathcal{V}|^k \times d}$, where we assume that the first k channels/modes of this tensor are indexed by the nodes of the graph. We use the notation $\mathbf{P} \star \mathcal{X}$ to denote the operation where we permute the first k channels of this tensor according the node permutation matrix $\mathbf{P}$. We can then define an linear equivariant layer as a linear operator (i.e., a tensor) $\mathcal{L} : \mathbb{R}^{|\mathcal{V}|^{k_1} \times d_1} \to \mathbb{R}^{|\mathcal{V}|^{k_2} \times d_2}$:

$$\mathcal{L} \times (\mathbf{P} \star \mathcal{X}) = \mathbf{P} \star (\mathcal{L} \times \mathcal{X}), \forall \mathbf{P} \in \mathcal{P}, \qquad (7.71)$$

where we use $\times$ to denote a generalized tensor product. Invariant linear operators can be similarly defined as tensors $\mathcal{L}$ that satisfy the following equality:

$$\mathcal{L} \times (\mathbf{P} \star \mathcal{X}) = \mathcal{L} \times \mathcal{X}, \forall \mathbf{P} \in \mathcal{P}. \qquad (7.72)$$

Note that both equivariant and invariant linear operators can be represented as tensors, but they have different structure. In particular, an equivariant operator $\mathcal{L} : \mathbb{R}^{|\mathcal{V}|^k \times d_1} \to \mathbb{R}^{|\mathcal{V}|^k \times d_2}$ corresponds to a tensor $\mathcal{L} \in \mathbb{R}^{|\mathcal{V}|^{2k} \times d_1 \times d_2}$, which has $2k$ channels indexed by nodes (i.e., twice as many node channels as the input). On the other hand, an invariant operator $\mathcal{L} : \mathbb{R}^{|\mathcal{V}|^k \times d_1} \to \mathbb{R}^{d_2}$ corresponds to a tensor $\mathcal{L} \in \mathbb{R}^{|\mathcal{V}|^k \times d_1 \times d_2}$, which has k channels indexed by nodes (i.e., the same number as the input). Interestingly, taking this tensor view of the linear operators, the equivariant (Equation (7.71)) and invariant (Equation (7.72)) properties for can be combined into a single requirement that the $\mathcal{L}$ tensor is a fixed point under node permutations:

$$\mathbf{P} \star \mathcal{L} = \mathcal{L}, \forall \mathbf{P} \in \mathcal{P}. \qquad (7.73)$$

In other words, for a given input $\mathcal{X} \in \mathbb{R}^{|V|^k \times d}$, both equivariant and invariant linear operators on this input will correspond to tensors that satisfy the fixed point in Equation (7.73), but the number of channels in the tensor will differ depending on whether it is an equivariant or invariant operator.

Maron et al. [2019] show that tensors satisfying the fixed point in Equation (7.73) can be constructed as a linear combination of a set of fixed basis elements. In particular, any order-l tensor $\mathcal{L}$ that satisfies Equation (7.73) can be written as

$$\mathcal{L} = \beta_1 \mathcal{B}_1 + \beta_2 + \ldots + \beta_{b(l)} \mathcal{B}_{b(l)}, \qquad (7.74)$$

where $\mathcal{B}_i$ are a set of fixed basis tensors, $\beta_i \in \mathbb{R}$ are real-valued weights, and $b(l)$ is the lth *Bell number*. The construction and derivation of these basis tensors is mathematically involved and is closely related to the theory of Bell numbers from combinatorics. However, a key fact and challenge is that the number of basis tensors needed grows with lth Bell number, which is an exponentially increasing series.

Using these linear equivariant and invariant layers, Maron et al. [2019] define their invariant k-order GNN model based on the following composition of functions:

$$\text{MLP} \circ \mathcal{L}_0 \circ \sigma \circ \mathcal{L}_1 \circ \sigma \mathcal{L}_2 \cdots \sigma \circ \mathcal{L}_m \times \mathcal{X}. \tag{7.75}$$

In this composition, we apply m equivariant linear layers $\mathcal{L}_1, ..., \mathcal{L}_m$, where $L_i : \mathcal{L} : \mathbb{R}^{|\mathcal{V}|^{k_i} \times d_1} \rightarrow \mathbb{R}^{|\mathcal{V}|^{k_i+1} \times d_2}$ with $\max_i k_i = k$ and $k_1 = 2$. Between each of these linear equivariant layers an element-wise nonlinearity, denoted by σ, is applied. The penultimate function in the composition, is an invariant linear layer, $\mathcal{L}_0$, which is followed by an MLP as the final function in the composition. The input to the k-order invariant GNN is the tensor $\mathcal{X} \in \mathbb{R}^{|\mathcal{V}|^2 \times d}$, where the first two channels correspond to the adjacency matrix and the remaining channels encode the initial node features/labels.

This approach is called k-order because the equivariant linear layers involve tensors that have up to k different channels. Most importantly, however, Maron et al. [2019] prove that k-order models following Equation (7.75) are equally powerful as the k-WL algorithm. As with the k-GNNs discussed in the previous section, however, constructing k-order invariant models for $k > 3$ is generally computationally intractable.

PART III

Generative Graph Models

CHAPTER 8

Traditional Graph Generation Approaches

The previous parts of this book introduced a wide variety of methods for learning representations of graphs. In this final part of the book, we will discuss a distinct but closely related task: the problem of *graph generation*.

The goal of graph generation is to build models that can generate realistic graph structures. In some ways, we can view this graph generation problem as the mirror image of the graph embedding problem. Instead of assuming that we are given a graph structure $\mathcal{G} = (\mathcal{V}, \mathcal{E})$ as *input* to our model, in graph generation we want the *output* of our model to be a graph $\mathcal{G}$. Of course, simply generating an arbitrary graph is not necessarily that challenging. For instance, it is trivial to generate a fully connected graph or a graph with no edges. The key challenge in graph generation, however, is generating graphs that have certain desirable properties. As we will see in the following chapters, the way in which we define "desirable properties"—and how we perform graph generation—varies drastically between different approaches.

In this chapter, we begin with a discussion of traditional approaches to graph generation. These traditional approaches pre-date most research on graph representation learning—and even machine learning research in general. The methods we will discuss in this chapter thus provide the backdrop to motivate the deep learning-based approaches that we will introduce in Chapter 9.

8.1 OVERVIEW OF TRADITIONAL APPROACHES

Traditional approaches to graph generation generally involve specifying some kind of *generative process*, which defines how the edges in a graph are created. In most cases we can frame this generative process as a way of specifying the probability or likelihood $P(\mathbf{A}[u, v] = 1)$ of an edge existing between two nodes u and v. The challenge is crafting some sort of generative process that is both tractable and also able to generate graphs with non-trivial properties or characteristics. Tractability is essential because we want to be able to sample or analyze the graphs that are generated. However, we also want these graphs to have some properties that make them good models for the kinds of graphs we see in the real world.

The three approaches we review in this subsection represent a small but representative subset of the traditional graph generation approaches that exist in the literature. For a more thorough survey and discussion, we recommend Newman [2018] as a useful resource.

8.2   ERDÖS–RÉNYI MODEL

Perhaps the simplest and most well-known generative model of graphs is the *Erdös–Rényi (ER)* model [Erdös and Rényi, 1960]. In this model we define the likelihood of an edge occurring between any pair of nodes as

$$P(\mathbf{A}[u, v] = 1) = r, \forall u, v \in \mathcal{V}, u \neq v, \tag{8.1}$$

where $r \in [0, 1]$ is parameter controlling the density of the graph. In other words, the ER model simply assumes that the probability of an edge occurring between any pairs of nodes is equal to r.

The ER model is attractive due to its simplicity. To generate a random ER graph, we simply choose (or sample) how many nodes we want, set the density parameter r, and then use Equation (8.1) to generate the adjacency matrix. Since the edge probabilities are all independent, the time complexity to generate a graph is $O(|\mathcal{V}|^2)$, i.e., linear in the size of the adjacency matrix.

The downside of the ER model, however, is that it does not generate very realistic graphs. In particular, the only property that we can control in the ER model is the density of the graph, since the parameter r is equal (in expectation) to the average degree in the graph. Other graph properties—such as the degree distribution, existence of community structures, node clustering coefficients, and the occurrence of structural motifs—are not captured by the ER model. It is well known that graphs generated by the ER model fail to reflect the distribution of these more complex graph properties, which are known to be important in the structure and function of real-world graphs.

8.3   STOCHASTIC BLOCK MODELS

Many traditional graph generation approaches seek to improve the ER model by better capturing additional properties of real-world graphs, which the ER model ignores. One prominent example is the class of *stochastic block models (SBMs)*, which seek to generate graphs with community structure.

In a basic SBM model, we specify a number γ of different blocks: $\mathcal{C}_1, ..., \mathcal{C}_\gamma$. Every node $u \in \mathcal{V}$ then has a probability p_i of belonging to block i, i.e., $p_i = P(u \in \mathcal{C}_i), \forall u \in \mathcal{V}, i = 1, ..., \gamma$ where $\sum_{i=1}^{\gamma} p_i = 1$. Edge probabilities are then specified by a block-to-block probability matrix $\mathbf{P} \in [0, 1]^{\gamma \times \gamma}$, where $\mathbf{C}[i, j]$ gives the probability of an edge occuring between a node in block $\mathcal{C}_i$ and a node in block $\mathcal{C}_j$. The generative process for the basic SBM model is as follows:

1. For every node $u \in \mathcal{V}$, we assign u to a block $\mathcal{C}_i$ by sampling from the categorical distribution defined by $(p_i, ..., p_\gamma)$.

2. For every pair of nodes $u \in C_i$ and $v \in C_j$ we sample an edge according to

$$P(\mathbf{A}[u, v] = 1) \propto \mathbf{C}[i, j]. \tag{8.2}$$

The key innovation in the SBM is that we can control the edge probabilities within and between different blocks, and this allows us to generate graphs that exhibit community structure. For example, a common SBM practice is to set a constant value α on the diagonal of the $\mathbf{C}$ matrix—i.e., $\mathbf{C}[i, i] = \alpha, i = 1, ..., \gamma$—and a separate constant $\beta < \alpha$ on the off-diagonal entries—i.e., $\mathbf{C}[i, j] = \beta, i, j = 1, ..\gamma, i \neq j$. In this setting, nodes have a probability α of having an edge with another node that assigned to the *same community* and a smaller probability $\beta < \alpha$ of having en edge with another node that is assigned to a *different community*.

The SBM model described above represents only the most basic variation of the general SBM framework. There are many variations of the SBM framework, including approaches for bipartite graphs, graphs with node features, as well as approaches to infer SBM parameters from data [Newman, 2018]. The key insight that is shared across all these approaches, however, is the idea of crafting a generative graph model that can capture the notion of communities in a graph.

8.4   PREFERENTIAL ATTACHMENT

The SBM framework described in the previous section can generate graphs with community structures. However, like the simple ER model, the SBM approach is limited in that it fails to capture the structural characteristics of individual nodes that are present in most real-world graphs.

For instance, in an SBM model, all nodes within a community have the same degree distribution. This means that the structure of individual communities is relatively homogeneous in that all the nodes have similar structural properties (e.g., similar degrees and clustering coefficients). Unfortunately, however, this homogeneity is quite unrealistic in the real world. In real-world graphs we often see much more heterogeneous and varied degree distributions, for example, with many low-degree nodes and a small number of high-degree "hub" nodes.

The third generative model we will introduce—termed the preferential attachment (PA) model—attempts to capture this characteristic property of real-world degree distributions [Albert and Barabási, 2002]. The PA model is built around the assumption that many real-world graphs exhibit *power law* degree distributions, meaning that the probability of a node u having degree d_u is roughly given by the following equation:

$$P(d_u = k) \propto k^{-\alpha}, \tag{8.3}$$

where $\alpha > 1$ is a parameter. Power law distributions—and other related distributions—have the property that they are *heavy tailed*. Formally, being heavy tailed means that a probability distribution goes to zero for extreme values slower than an exponential distribution. This means that heavy-tailed distributions assign non-trivial probability mass to events that are essentially "impossible" under a standard exponential distribution. In the case of degree distributions, this heavy tailed nature essentially means that there is a non-zero chance of encountering a small number of very high-degree nodes. Intuitively, power law degree distributions capture the fact that real-world graphs have a large number of nodes with small degrees but also have a small number of nodes with extremely large degrees.[1]

The PA model generates graphs that exhibit power-law degree distributions using a simple generative process:

1. First, we initialize a fully connected graph with m_0 nodes.

2. Next, we iteratively add $n - m_0$ nodes to this graph. For each new node u that we add at iteration t, we connect it to $m < m_0$ existing nodes in the graph, and we choose its m neighbors by sampling without replacement according to the following probability distribution:

$$P(\mathbf{A}[u, v]) = \frac{d_v^{(t)}}{\sum_{v' \in \mathcal{V}^{(t)}} d_{v'}^{(t)}}, \tag{8.4}$$

where $d_v^{(t)}$ denotes the degree of node v at iteration t and $\mathcal{V}^{(t)}$ denotes the set of nodes that have been added to the graph up to iteration t.

The key idea is that the PA model connects new nodes to existing nodes with a probability that is proportional to the existing nodes' degrees. This means that high degree nodes will tend to accumulate more and more neighbors in a *rich get richer* phenomenon as the graph grows. One can show that the PA model described above generates connected graphs that have power law degree distributions with $\alpha = 3$ [Albert and Barabási, 2002].

An important aspect of the PA model—which distinguishes it from the ER and SBM models—is that the generation process is *autoregressive*. Instead of specifying the edge probabilities for the entire graph in one step, the PA model relies on an iterative approach, where the edge probabilities at step t depend on the edges that were added at step $t - 1$. We will see that this notion of autoregressive generation will reoccur within the context of deep learning approaches to graph generation in Chapter 8.

8.5   TRADITIONAL APPLICATIONS

The three previous subsections outlined three traditional graph generation approaches: the ER model, the SBM, and the PA model. The insight in these models is that we specify a generation process or probability model, which allows us to capture some useful property of real-world graphs while still being tractable and easy to analyze. Historically, these traditional generation models have been used in two key applications.

---

[1]There is a great deal of controversy regarding the prevalence of actual power law distributions in real-world data. There is compelling evidence that many supposedly power-law distributions are in fact better modeled by distributions like the log-normal. Clauset et al. [2009] contains a useful discussion and empirical analysis of this issue.

Generating Synthetic Data for Benchmarking and Analysis Tasks

The first useful application of these generative models is that they can be used to generate synthetic graphs for benchmarking and analysis tasks. For example, suppose you've developed a community detection algorithm. It would be reasonable to expect that your community detection approach should be able to infer the underlying communities in a graph generated by an SBM model. Similarly, if you have designed a network analysis engine that is suppose to scale to very large graphs, it would be good practice to test your framework on synthetic graphs generated by the PA model, in order to ensure your analysis engine can handle heavy-tailed degree distributions.

Creating Null Models

The second key application task for traditional graph generation approaches is the creation of null models. Suppose you are researching a social network dataset. After analyzing this network and computing various statistics—such as degree distributions and clustering coefficients—you might want to ask the following question: How surprising are the characteristics of this network? Generative graph models provide a precise way for us to answer this question. In particular, we can investigate the extent to which different graph characteristics are probable (or unexpected) under different generative models. For example, the presence of heavy-tailed degree distributions in a social network might seem surprising at first glance, but this property is actually expected if we assume that the data is generated according to a preferential attachment process. In general, traditional generative models of graphs give us the ability to interrogate what sorts of graph characteristics can be easily explained by simple generative processes. In a statistical sense, they provide us with *null models* that we can use as reference points for our understanding of real-world graphs.

CHAPTER 9

Deep Generative Models

The traditional graph generation approaches discussed in the previous chapter are useful in many settings. They can be used to efficiently generate synthetic graphs that have certain properties, and they can be used to give us insight into how certain graph structures might arise in the real world. However, a key limitation of those traditional approaches is that they rely on a fixed, hand-crafted generation process. In short, the traditional approaches can generate graphs, but they lack the ability to *learn* a generative model from data.

In this chapter, we will introduce various approaches that address exactly this challenge. These approaches will seek to learn a generative model of graphs based on a set of *training graphs*. These approaches avoid hand-coding particular properties—such as community structure or degree distributions—into a generative model. Instead, the goal of these approaches is to design models that can observe a set of graphs $\{\mathcal{G}_1, ..., \mathcal{G}_n\}$ and learn to generate graphs with similar characteristics as this training set.

We will introduce a series of basic deep generative models for graphs. These models will adapt three of the most popular approaches to building general deep generative models: variational autoencoders (VAEs), generative adversarial networks (GANs), and autoregressive models. We will focus on the simple and general variants of these models, emphasizing the high-level details and providing pointers to the literature where necessary. Moreover, while these generative techniques can in principle be combined with one another—for example, VAEs are often combined with autoregressive approaches—we will not discuss such combinations in detail here. Instead, we will begin with a discussion of basic VAE models for graphs, where we seek to generate an entire graph *all-at-once* in an autoencoder style. Following this, we will discuss how GAN-based objectives can be used in lieu of variational losses, but still in the setting where the graphs are generated all-at-once. These all-at-once generative models are analogous to the ER and SBM generative models from the last chapter, in that we sample all edges in the graph simultaneously. Finally, the chapter will close with a discussion of autoregressive approaches, which allow one to generate a graph *incrementally* instead of all-at-once (e.g., generating a graph node-by-node). These autoregressive approaches bear similarities to the preferential attachment model from the previous chapter in that the probability of adding an edge at each step during generation depends on what edges were previously added to the graph.

For simplicity, all the methods we discuss will only focus on generating graph structures (i.e., adjacency matrices) and not on generating node or edge features. This chapter assumes a basic familiarity with VAEs, GANs, and autoregressive generative models, such as LSTM-

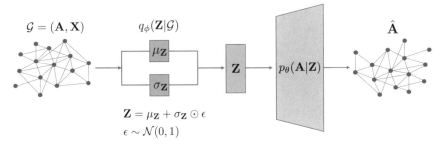

Figure 9.1: Illustration of a standard VAE model applied to the graph setting. An *encoder* neural network maps the input graph $\mathcal{G} = (\mathbf{A}, \mathbf{X})$ to a *posterior distribution* $q_\phi(\mathbf{Z}|\mathcal{G})$ over latent variables $\mathbf{Z}$. Given a sample from this posterior, the *decoder* model $p_\theta(\mathbf{A}|\mathbf{Z})$ attempts to reconstruct the adjacency matrix.

based language models. We refer the reader to Goodfellow et al. [2016] for background reading in these areas.

Of all the topics in this book, deep generative models of graphs are both the most technically involved and the most nascent in their development. Thus, our goal in this chapter is to introduce the key methodological frameworks that have inspired the early research in this area, while also highlighting a few influential models. As a consequence, we will often eschew low-level details in favor of a more high-level tour of this nascent sub-area.

9.1 VARIATIONAL AUTOENCODER APPROACHES

VAEs are one of the most popular approaches to develop deep generative models [Kingma and Welling, 2013]. The theory and motivation of VAEs is deeply rooted in the statistical domain of variational inference, which we briefly touched upon in Chapter 7. However, for the purposes of this book, the key idea behind applying a VAE to graphs can be summarized as follows (Figure 9.1): our goal is to train a *probabilistic decoder* model $p_\theta(\mathbf{A}|\mathbf{Z})$, from which we can sample realistic graphs (i.e., adjacency matrices) $\hat{\mathbf{A}} \sim p_\theta(\mathbf{A}|\mathbf{Z})$ by conditioning on a latent variable $\mathbf{Z}$. In a probabilistic sense, we aim to learn a conditional distribution over adjacency matrices (with the distribution begin conditioned on some latent variable).

In order to train a VAE, we combine the probabilistic decoder with a *probabilistic encoder* model $q_\theta(\mathbf{Z}|\mathcal{G})$. This encoder model maps an input graph $\mathcal{G}$ to a *posterior distribution* over the latent variable $\mathbf{Z}$. The idea is that we jointly train the encoder and decoder so that the decoder is able to reconstruct training graphs given a latent variable $\mathbf{Z} \sim q_\theta(\mathbf{Z}|\mathcal{G})$ sampled from the encoder. Then, after training, we can discard the encoder and generate new graphs by sampling latent variables $\mathbf{Z} \sim p(\mathbf{Z})$ from some (unconditional) prior distribution and feeding these sampled latents to the decoder.

In more formal and mathematical detail, to build a VAE for graphs we must specify the following key components.

1. A *probabilistic encoder* model q_ϕ. In the case of graphs, the probabilistic encoder model takes a graph $\mathcal{G}$ as input. From this input, q_ϕ then defines a distribution $q_\phi(\mathbf{Z}|\mathcal{G})$ over *latent representations*. Generally, in VAEs the *reparameterization trick* with Gaussian random variables is used to design a probabilistic q_ϕ function. That is, we specify the latent conditional distribution as $\mathbf{Z} \sim \mathcal{N}(\mu_\phi(\mathcal{G}), \sigma(\phi(\mathcal{G})))$, where μ_ϕ and σ_ϕ are neural networks that generate the mean and variance parameters for a normal distribution, from which we sample latent embeddings $\mathbf{Z}$.

2. A *probabilistic decoder* model p_θ. The decoder takes a latent representation $\mathbf{Z}$ as input and uses this input to specify a conditional distribution over graphs. In this chapter, we will assume that p_θ defines a conditional distribution over the entries of the adjacency matrix, i.e., we can compute $p_\theta(\mathbf{A}[u, v] = 1|\mathbf{Z})$.

3. A *prior distribution* $p(\mathbf{Z})$ over the latent space. In this chapter we will adopt the standard Gaussian prior $\mathbf{Z} \sim \mathcal{N}(\mathbf{0}, \mathbf{1})$, which is commonly used for VAEs.

Given these components and a set of training graphs $\{\mathcal{G}_1, .., \mathcal{G}_n\}$, we can train a VAE model by minimizing the evidence likelihood lower bound (ELBO):

$$\mathcal{L} = \sum_{\mathcal{G}_i \in \{G_1, ..., \mathcal{G}_n\}} \mathbb{E}_{q_\theta(\mathbf{z}|\mathcal{G}_i)}[p_\theta(\mathcal{G}_i|\mathbf{Z})] - \mathrm{KL}(q_\theta(\mathbf{Z}|\mathcal{G}_i) \| p(\mathbf{Z})). \tag{9.1}$$

The basic idea is that we seek to maximize the reconstruction ability of our decoder—i.e., the likelihood term $\mathbb{E}_{q_\theta(\mathbf{z}|\mathcal{G}_i)}[p_\theta(\mathcal{G}_i|\mathbf{Z})]$—while minimizing the KL-divergence between our posterior latent distribution $q_\theta(\mathbf{Z}|\mathcal{G}_i)$ and the prior $p(\mathbf{Z})$.

The motivation behind the ELBO loss function is rooted in the theory of variational inference [Wainwright and Jordan, 2008]. However, the key intuition is that we want to generate a distribution over latent representations so that the following two (conflicting) goals are satisfied.

1. The sampled latent representations encode enough information to allow our decoder to reconstruct the input.

2. The latent distribution is as close as possible to the prior.

The first goal ensures that we learn to decode meaningful graphs from the encoded latent representations, when we have training graphs as input. The second goal acts as a regularizer and ensures that we can decode meaningful graphs even when we sample latent representations from the prior $p(\mathbf{Z})$. This second goal is critically important if we want to generate new graphs after training: we can generate new graphs by sampling from the prior and feeding these latent embeddings to the decoder, and this process will only work if this second goal is satisfied.

In the following sections, we will describe two different ways in which the VAE idea can be instantiated for graphs. The approaches differ in how they define the encoder, decoder, and the latent representations. However, they share the overall idea of adapting the VAE model to graphs.

9.1.1 NODE-LEVEL LATENTS

The first approach we will examine builds closely upon the idea of encoding and decoding graphs based on node embeddings, which we introduced in Chapter 3. The key idea in this approach is that that the encoder generates latent representations for each node in the graph. The decoder then takes pairs of embeddings as input and uses these embeddings to predict the likelihood of an edge occurring between the two nodes. This idea was first proposed by Kipf and Welling [2016b] and termed the Variational Graph Autoencoder (VGAE).

Encoder Model

The encoder model in this setup can be based on any of the GNN architectures we discussed in Chapter 5. In particular, given an adjacency matrix $\mathbf{A}$ and node features $\mathbf{X}$ as input, we use two separate GNNs to generate mean and variance parameters, respectively, conditioned on this input:

$$\mu_{\mathbf{Z}} = \text{GNN}_\mu(\mathbf{A}, \mathbf{X}) \qquad \log \sigma_{\mathbf{Z}} = \text{GNN}_\sigma(\mathbf{A}, \mathbf{X}). \tag{9.2}$$

Here, $\mu_{\mathbf{Z}}$ is a $|\mathcal{V}| \times d$-dimensional matrix, which specifies a mean embedding value *for each node in the input graph*. The $\log \sigma_{\mathbf{Z}} \in \mathbb{R}^{|V| \times d}$ matrix similarly specifies the log-variance for the latent embedding of each node.[1]

Given the encoded $\mu_{\mathbf{Z}}$ and $\log \sigma_{\mathbf{Z}}$ parameters, we can sample a set of latent node embeddings by computing

$$\mathbf{Z} = \epsilon \circ \exp\left(\log(\sigma_{\mathbf{Z}})\right) + \mu_{\mathbf{Z}}, \tag{9.3}$$

where $\epsilon \sim \mathcal{N}(\mathbf{0}, \mathbf{1})$ is a $|\mathcal{V}| \times d$ dimensional matrix with independently sampled unit normal entries.

The Decoder Model

Given a matrix of sampled node embeddings $\mathbf{Z} \in \mathbb{R}^{|V| \times d}$, the goal of the decoder model is to predict the likelihood of all the edges in the graph. Formally, the decoder must specify $p_\theta(\mathbf{A}|\mathbf{Z})$— the posterior probability of the adjacency matrix given the node embeddings. Again, here, many of the techniques we have already discussed in this book can be employed, such as the various edge decoders introduced in Chapter 3. In the original VGAE paper, Kipf and Welling [2016b] employ a simple dot-product decoder defined as follows:

$$p_\theta(\mathbf{A}[u, v] = 1|\mathbf{z}_u, \mathbf{z}_v) = \sigma(\mathbf{z}_u^\top \mathbf{z}_v), \tag{9.4}$$

---

[1]Parameterizing the log-variance is often more stable than directly parameterizing the variance.

where σ is used to denote the sigmoid function. Note, however, that a variety of edge decoders could feasibly be employed, as long as these decoders generate valid probability values.

To compute the reconstruction loss in Equation (9.1) using this approach, we simply assume independence between edges and define the posterior $p_\theta(\mathcal{G}|\mathbf{Z})$ over the full graph as follows:

$$p_\theta(\mathcal{G}|\mathbf{Z}) = \prod_{(u,v)\in\mathcal{V}^2} p_\theta(\mathbf{A}[u,v] = 1|\mathbf{z}_u, \mathbf{z}_v), \qquad (9.5)$$

which corresponds to a binary cross-entropy loss over the edge probabilities. To generate discrete graphs after training, we can sample edges based on the posterior Bernoulli distributions in Equation (9.4).

Limitations

The basic VGAE model sketched in the previous sections defines a valid generative model for graphs. After training this model to reconstruct a set of training graphs, we could sample node embeddings $\mathbf{Z}$ from a standard normal distribution and use our decoder to generate a graph. However, the generative capacity of this basic approach is extremely limited, especially when a simple dot-product decoder is used. The main issue is that the decoder has no parameters, so the model is not able to generate non-trivial graph structures without a training graph as input. Indeed, in their initial work on the subject, Kipf and Welling [2016b] proposed the VGAE model as an approach to generate node embeddings, but they did not intend it as a generative model to sample new graphs.

Some papers have proposed to address the limitations of VGAE as a generative model by making the decoder more powerful. For example, Grover et al. [2019] propose to augment the decoder with an "iterative" GNN-based decoder. Nonetheless, the simple node-level VAE approach has not emerged as a successful and useful approach for graph generation. It has achieved strong results on reconstruction tasks and as an autoencoder framework, but as a generative model, this simple approach is severely limited.

9.1.2 GRAPH-LEVEL LATENTS

As an alternative to the node-level VGAE approach described in the previous section, one can also define variational autoencoders based on graph-level latent representations. In this approach, we again use the ELBO loss (Equation (9.1)) to train a VAE model. However, we modify the encoder and decoder functions to work with graph-level latent representations $\mathbf{z}_\mathcal{G}$. The graph-level VAE described in this section was first proposed by Simonovsky and Komodakis [2018], under the name GraphVAE.

Encoder Model

The encoder model in a graph-level VAE approach can be an arbitrary GNN model augmented with a pooling layer. In particular, we will let GNN : $\mathbb{Z}^{|\mathcal{V}|\times|\mathcal{V}|} \times \mathbb{R}^{|V|\times m} \to \mathbb{R}^{||V|\times d}$ denote any

k-layer GNN, which outputs a matrix of node embeddings, and we will use POOL : $\mathbb{R}^{||V|\times d} \rightarrow \mathbb{R}^d$ to denote a pooling function that maps a matrix of node embeddings $\mathbf{Z} \in \mathbb{R}^{|V|\times d}$ to a graph-level embedding vector $\mathbf{z}_{\mathcal{G}} \in \mathbb{R}^d$ (as described in Chapter 5). Using this notation, we can define the encoder for a graph-level VAE by the following equations:

$$\mu_{\mathbf{z}_{\mathcal{G}}} = \text{POOL}_\mu \left(\text{GNN}_\mu(\mathbf{A}, \mathbf{X}) \right) \qquad \log \sigma_{\mathbf{z}_{\mathcal{G}}} = \text{POOL}_\sigma \left(\text{GNN}_\sigma(\mathbf{A}, \mathbf{X}) \right), \qquad (9.6)$$

where again we use two separate GNNs to parameterize the mean and variance of a posterior normal distribution over latent variables. Note the critical difference between this graph-level encoder and the node-level encoder from the previous section: here, we are generating a mean $\mu_{\mathbf{z}_{\mathcal{G}}} \in \mathbb{R}^d$ and variance parameter $\log \sigma_{\mathbf{z}_{\mathcal{G}}} \in \mathbb{R}^d$ for a single graph-level embedding $\mathbf{z}_{\mathcal{G}} \sim \mathcal{N}(\mu_{\mathbf{z}_{\mathcal{G}}}, \sigma_{\mathbf{z}_{\mathcal{G}}})$, whereas in the previous section we defined posterior distributions for each individual node.

Decoder Model

The goal of the decoder model in a graph-level VAE is to define $p_\theta(\mathcal{G}|\mathbf{z}_{\mathcal{G}})$, the posterior distribution of a particular graph structure given the graph-level latent embedding. The original Graph-VAE model proposed to address this challenge by combining a basic MLP with a Bernoulli distributional assumption [Simonovsky and Komodakis, 2018]. In this approach, we use an MLP to map the latent vector $\mathbf{z}_{\mathcal{G}}$ to a matrix $\tilde{\mathbf{A}} \in [0, 1]^{|\mathcal{V}|\times|\mathcal{V}|}$ of edge probabilities:

$$\tilde{\mathbf{A}} = \sigma \left(\text{MLP}(\mathbf{z}_{\mathcal{G}}) \right), \qquad (9.7)$$

where the sigmoid function σ is used to guarantee entries in $[0, 1]$. In principle, we can then define the posterior distribution in an analogous way as the node-level case:

$$p_\theta(\mathcal{G}|\mathbf{z}_{\mathcal{G}}) = \prod_{(u,v)\in\mathcal{V}} \tilde{\mathbf{A}}[u, v]\mathbf{A}[u, v] + (1 - \tilde{\mathbf{A}}[u, v])(1 - \mathbf{A}[u, v]), \qquad (9.8)$$

where $\mathbf{A}$ denotes the true adjacency matrix of graph $\mathcal{G}$ and $\tilde{\mathbf{A}}$ is our predicted matrix of edge probabilities. In other words, we simply assume independent Bernoulli distributions for each edge, and the overall log-likelihood objective is equivalent to set of independent binary cross-entropy loss function on each edge. However, there are two key challenges in implementing Equation (9.8) in practice:

1. First, if we are using an MLP as a decoder, then **we need to assume a fixed number of nodes.** Generally, this problem is addressed by assuming a *maximum* number of nodes and using a *masking* approach. In particular, we can assume a maximum number of nodes $n_{\max}$, which limits the output dimension of the decoder MLP to matrices of size $n_{\max} \times n_{\max}$. To decode a graph with $|\mathcal{V}| < n_{\max}$ nodes during training, we simply mask (i.e., ignore) all entries in $\tilde{\mathbf{A}}$ with row or column indices greater than $|\mathcal{V}|$. To generate graphs of varying sizes after the model is trained, we can specify a distribution $p(n)$ over graph sizes

with support $\{2, ..., n_{\max}\}$ and sample from this distribution to determine the size of the generated graphs. A simple strategy to specify $p(n)$ is to use the empirical distribution of graph sizes in the training data.

2. The second key challenge in applying Equation (9.8) in practice is that **we do not know the correct ordering of the rows and columns in $\tilde{\mathbf{A}}$ when we are computing the reconstruction loss**. The matrix $\tilde{\mathbf{A}}$ is simply generated by an MLP, and when we want to use $\tilde{\mathbf{A}}$ to compute the likelihood of a training graph, we need to implicitly assume some ordering over the nodes (i.e., the rows and columns of $\tilde{\mathbf{A}}$). Formally, the loss in Equation (9.8) requires that we specify a node ordering $\pi \in \Pi$ to order the rows and columns in $\tilde{\mathbf{A}}$.

This is important because if we simply ignore this issue then the decoder can overfit to the arbitrary node orderings used during training. There are two popular strategies to address this issue. The first approach—proposed by Simonovsky and Komodakis [2018]—is to apply a graph matching heuristic to try to find the node ordering of $\tilde{\mathbf{A}}$ for each training graph that gives the highest likelihood, which modifies the loss to

$$p_\theta(\mathcal{G}|\mathbf{z}_\mathcal{G}) = \max_{\pi \in \Pi} \prod_{(u,v)\in\mathcal{V}} \tilde{\mathbf{A}}^\pi[u,v]\mathbf{A}[u,v] + (1 - \tilde{\mathbf{A}}^\pi[u,v])(1 - \mathbf{A}[u,v]), \qquad (9.9)$$

where we use $\tilde{\mathbf{A}}^\pi$ to denote the predicted adjacency matrix under a specific node ordering π. Unfortunately, however, computing the maximum in Equation (9.9)—even using heuristic approximations—is computationally expensive, and models based on graph matching are unable to scale to graphs with more than hundreds of nodes. More recently, authors have tended to use heuristic node orderings. For example, we can order nodes based on a depth-first or breadth-first search starting from the highest-degree node. In this approach, we simply specify a particular ordering function π and compute the loss with this ordering:

$$p_\theta(\mathcal{G}|\mathbf{z}_\mathcal{G}) \approx \prod_{(u,v)\in\mathcal{V}} \tilde{\mathbf{A}}^\pi[u,v]\mathbf{A}[u,v] + (1 - \tilde{\mathbf{A}}^\pi[u,v])(1 - \mathbf{A}[u,v]),$$

or we consider a small set of heuristic orderings $\pi_1, ..., \pi_n$ and average over these orderings:

$$p_\theta(\mathcal{G}|\mathbf{z}_\mathcal{G}) \approx \sum_{\pi_i \in \{\pi_1,...,\pi_n\}} \prod_{(u,v)\in\mathcal{V}} \tilde{\mathbf{A}}^{\pi_i}[u,v]\mathbf{A}[u,v] + (1 - \tilde{\mathbf{A}}^{\pi_i}[u,v])(1 - \mathbf{A}[u,v]).$$

These heuristic orderings do not solve the graph matching problem, but they seem to work well in practice. Liao et al. [2019a] provides a detailed discussion and comparison of these heuristic ordering approaches, as well as an interpretation of this strategy as a variational approximation.

Limitations

As with the node-level VAE approach, the basic graph-level framework has serious limitations. Most prominently, using graph-level latent representations introduces the issue of specifying node orderings, as discussed above. This issue—together with the use of MLP decoders—currently limits the application of the basic graph-level VAE to small graphs with hundreds of nodes or less. However, the graph-level VAE framework can be combined with more effective decoders—including some of the autoregressive methods we discuss in Section 9.3—which can lead to stronger models. We will mention one prominent example of such as approach in Section 9.5, when we highlight the specific task of generating molecule graph structures.

9.2   ADVERSARIAL APPROACHES

VAEs are a popular framework for deep generative models—not just for graphs, but for images, text, and a wide-variety of data domains. VAEs have a well-defined probabilistic motivation, and there are many works that leverage and analyze the structure of the latent spaces learned by VAE models. However, VAEs are also known to suffer from serious limitations—such as the tendency for VAEs to produce blurry outputs in the image domain. Many recent state-of-the-art generative models leverage alternative generative frameworks, with GANs being one of the most popular [Goodfellow et al., 2014].

The basic idea behind a general GAN-based generative models is as follows. First, we define a trainable generator network $g_\theta : \mathbb{R}^d \to \mathcal{X}$. This generator network is trained to generate realistic (but fake) data samples $\tilde{\mathbf{x}} \in \mathcal{X}$ by taking a random seed $\mathbf{z} \in \mathbb{R}^d$ as input (e.g., a sample from a normal distribution). At the same time, we define a discriminator network $d_\phi : \mathcal{X} \to [0, 1]$. The goal of the discriminator is to distinguish between real data samples $\mathbf{x} \in \mathcal{X}$ and samples generated by the generator $\tilde{\mathbf{x}} \in \mathcal{X}$. Here, we will assume that discriminator outputs the probability that a given input is fake.

To train a GAN, both the generator and discriminator are optimized jointly in an *adversarial game*:

$$\min_{\theta} \max_{\phi} \mathbb{E}_{\mathbf{x} \sim p_{\mathrm{data}}(\mathbf{x})}[\log(1 - d_\phi(\mathbf{x}))] + \mathbb{E}_{\mathbf{z} \sim p_{\mathrm{seed}}(\mathbf{z})}[\log(d_\phi(g_\theta(\mathbf{z}))], \tag{9.10}$$

where $p_{\mathrm{data}}(\mathbf{x})$ denotes the empirical distribution of real data samples (e.g., a uniform sample over a training set) and $p_{\mathrm{seed}}(\mathbf{z})$ denotes the random seed distribution (e.g., a standard multivariate normal distribution). Equation (9.10) represents a minimax optimization problem. The generator is attempting to minimize the discriminatory power of the discriminator, while the discriminator is attempting to maximize its ability to detect fake samples. The optimization of the GAN minimax objective—as well as more recent variations—is challenging, but there is a wealth of literature emerging on this subject [Brock et al., 2018, Heusel et al., 2017, Mescheder et al., 2018].

A Basic GAN Approach to Graph Generation

In the context of graph generation, a GAN-based approach was first employed in concurrent work by Bojchevski et al. [2018] and De Cao and Kipf [2018]. The basic approach proposed by De Cao and Kipf [2018]—which we focus on here—is similar to the graph-level VAE discussed in the previous section. For instance, for the generator, we can employ a simple multi-layer perceptron (MLP) to generate a matrix of edge probabilities given a seed vector $\mathbf{z}$:

$$\tilde{\mathbf{A}} = \sigma\left(\mathrm{MLP}(\mathbf{z})\right). \tag{9.11}$$

Given this matrix of edge probabilities, we can then generate a discrete adjacency matrix $\hat{\mathbf{A}} \in \mathbb{Z}^{|\mathcal{V}| \times |\mathcal{V}|}$ by sampling independent Bernoulli variables for each edge, with probabilities given by the entries of $\tilde{\mathbf{A}}$, i.e., $\hat{\mathbf{A}}[u, v] \sim \mathrm{Bernoulli}(\tilde{\mathbf{A}}[u, v])$. For the discriminator, we can employ any GNN-based graph classification model. The generator model and the discriminator model can then be trained according to Equation (9.10) using standard tools for GAN optimization.

Benefits and Limitations of the GAN Approach

As with the VAE approaches, the GAN framework for graph generation can be extended in various ways. More powerful generator models can be employed—for instance, leveraging the autoregressive techniques discussed in the next section—and one can even incorporate node features into the generator and discriminator models [De Cao and Kipf, 2018].

One important benefit of the GAN-based framework is that it removes the complication of specifying a node ordering in the loss computation. As long as the discriminator model is permutation invariant—which is the case for almost every GNN—then the GAN approach does not require any node ordering to be specified. The ordering of the adjacency matrix generated by the generator is immaterial if the discriminator is permutation invariant. However, despite this important benefit, GAN-based approaches to graph generation have so far received less attention and success than their variational counterparts. This is likely due to the difficulties involved in the minimax optimization that GAN-based approaches require, and investigating the limits of GAN-based graph generation is currently an open problem.

9.3   AUTOREGRESSIVE METHODS

The previous two sections detailed how the ideas of VAEs and GANs can be applied to graph generation. However, both the basic GAN and VAE-based approaches that we discussed used simple MLPs to generate adjacency matrices. In this section, we will introduce more sophisticated *autoregressive* methods that can decode graph structures from latent representations. The methods that we introduce in this section can be combined with the GAN and VAE frameworks that we introduced previously, but they can also be trained as standalone generative models.

9.3.1   MODELING EDGE DEPENDENCIES

The simple generative models discussed in the previous sections assumed that edges were generated *independently*. From a probabilistic perspective, we defined the likelihood of a graph given a latent representation $\mathbf{z}$ by decomposing the overall likelihood into a set of independent edge likelihoods as follows:

$$P(\mathcal{G}|\mathbf{z}) = \prod_{(u,v)\in\mathcal{V}^2} P(\mathbf{A}[u,v]|\mathbf{z}). \tag{9.12}$$

Assuming independence between edges is convenient, as it simplifies the likelihood model and allows for efficient computations. However, it is a strong and limiting assumption, since real-world graphs exhibit many complex dependencies between edges. For example, the tendency for real-world graphs to have high clustering coefficients is difficult to capture in an edge-independent model. To alleviate this issue—while still maintaining tractability—autoregressive model relax the assumption of edge independence.

Instead, in the autoregressive approach, we assume that edges are generated sequentially and that the likelihood of each edge can be conditioned on the edges that have been previously generated. To make this idea precise, we will use $\mathbf{L}$ to denote the lower-triangular portion of the adjacency matrix $\mathbf{A}$. Assuming we are working with simple graphs, $\mathbf{A}$ and $\mathbf{L}$ contain exactly the same information, but it will be convenient to work with $\mathbf{L}$ in the following equations. We will then use the notation $\mathbf{L}[v_1,:]$ to denote the row of $\mathbf{L}$ corresponding to node v_1, and we will assume that the rows of $\mathbf{L}$ are indexed by nodes $v_1, ..., v_{|\mathcal{V}|}$. Note that due to the lower-triangular nature of $\mathbf{L}$, we will have that $\mathbf{L}[v_i, v_j] = 0, \forall j > i$, meaning that we only need to be concerned with generating the first i entries for any row $\mathbf{L}[v_i,:]$; the rest can simply be padded with zeros. Given this notation, the autoregressive approach amounts to the following decomposition of the overall graph likelihood:

$$P(\mathcal{G}|\mathbf{z}) = \prod_{i=1}^{|\mathcal{V}|} P(\mathbf{L}[v_i,:]|\mathbf{L}[v_1,:], ..., \mathbf{L}[v_{i-1},:], \mathbf{z}). \tag{9.13}$$

In other words, when we generate row $\mathbf{L}[v_i,:]$, we condition on all the previous generated rows $\mathbf{L}[v_j,:]$ with $j < i$.

9.3.2   RECURRENT MODELS FOR GRAPH GENERATION

We will now discuss two concrete instantiations of the autoregressive generation idea. These two approaches build upon ideas first proposed in Li et al. [2018] and are generally indicative of the strategies that one could employ for this task. In the first model we will review—called GraphRNN [You et al., 2018]—we model autoregressive dependencies using an RNN. In the second approach—called graph recurrent attention network (GRAN) [Liao et al., 2019a]—we generate graphs by using a GNN to condition on the adjacency matrix that has been generated so far.

GraphRNN

The first model to employ this autoregressive generation approach was GraphRNN [You et al., 2018]. The basic idea in the GraphRNN approach is to use a hierarchical RNN to model the edge dependencies in Equation (9.13).

The first RNN in the hierarchical model—termed the graph-level RNN—is used to model the state of the graph that has been generated so far. Formally, the graph-level RNN maintains a hidden state $\mathbf{h}_i$, which is updated after generating each row of the adjacency matrix $\mathbf{L}[v_i, :]$:

$$\mathbf{h}_{i+1} = \text{RNN}_{\text{graph}}(\mathbf{h}_i, \mathbf{L}[v_i, L]), \tag{9.14}$$

where we use $\text{RNN}_{\text{graph}}$ to denote a generic RNN state update with $\mathbf{h}_i$ corresponding to the hidden state and $\mathbf{L}[v_i, L]$ to the observation.[2] In You et al. [2018]'s original formulation, a fixed initial hidden state $\mathbf{h}_0 = \mathbf{0}$ is used to initialize the graph-level RNN, but in principle this initial hidden state could also be learned by a graph encoder model or sampled from a latent space in a VAE-style approach.

The second RNN—termed the node-level RNN or RNN_{node}—generates the entries of $\mathbf{L}[v_i, :]$ in an autoregressive manner. RNN_{node} takes the graph-level hidden state $\mathbf{h}_i$ as an initial input and then sequentially generates the binary values of $\mathbf{L}[v_i, :]$, assuming a conditional Bernoulli distribution for each entry. The overall GraphRNN approach is called hierarchical because the node-level RNN is initialized at each time-step with the current hidden state of the graph-level RNN.

Both the graph-level $\text{RNN}_{\text{graph}}$ and the node-level RNN_{node} can be optimized to maximize the likelihood the training graphs (Equation (9.13)) using the *teaching forcing* strategy [Williams and Zipser, 1989], meaning that the ground truth values of $\mathbf{L}$ are always used to update the RNNs during training. To control the size of the generated graphs, the RNNs are also trained to output end-of-sequence tokens, which are used to specify the end of the generation process. Note that—as with the graph-level VAE approaches discussed in Section 9.1—computing the likelihood in Equation (9.13) requires that we assume a particular ordering over the generated nodes.

After training to maximize the likelihood of the training graphs (Equation (9.13)), the GraphRNN model can be used to generate graphs at test time by simply running the hierarchical RNN starting from the fixed, initial hidden state $\mathbf{h}_0$. Since the edge-level RNN involves a stochastic sampling process to generate the discrete edges, the GraphRNN model is able to generate diverse samples of graphs even when a fixed initial embedding is used. However—as mentioned above—the GraphRNN model could, in principle, be used as a decoder or generator within a VAE or GAN framework, respectively.

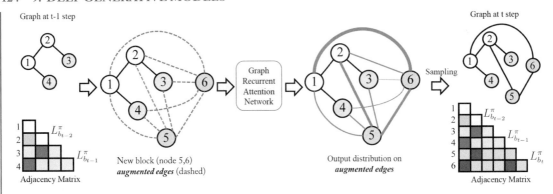

Figure 9.2: Illustration of the GRAN generation approach [Liao et al., 2019a].

Graph Recurrent Attention Networks (GRAN)

The key benefit of the GraphRNN approach—discussed above—is that it models dependencies between edges. Using an autoregressive modeling assumption (Equation (9.13)), GraphRNN can condition the generation of edges at generation step i based on the state of the graph that has already been generated during generation steps $1, \dots i-1$. Conditioning in this way makes it much easier to generate complex motifs and regular graph structures, such as grids. For example, in Figure 9.3, we can see that GraphRNN is more capable of generating grid-like structures, compared to the basic graph-level VAE (Section 9.1). However, the GraphRNN approach still has serious limitations. As we can see in Figure 9.3, the GraphRNN model still generates unrealistic artifacts (e.g., long chains) when trained on samples of grids. Moreover, GraphRNN can be difficult to train and scale to large graphs due to the need to backpropagate through many steps of RNN recurrence.

To address some of the limitations of the GraphRNN approach, Liao et al. [2019a] proposed the GRAN model. GRAN—which stands for *graph recurrent attention networks*—maintains the autoregressive decomposition of the generation process. However, instead of using RNNs to model the autoregressive generation process, GRAN uses GNNs. The key idea in GRAN is that we can model the conditional distribution of each row of the adjacency matrix by running a GNN on the graph that has been generated so far (Figure 9.2):

$$P(\mathbf{L}[v_i,:]|\mathbf{L}[v_1,:],\dots,\mathbf{L}[v_{i-1},:],\mathbf{z}) \approx \text{GNN}(\mathbf{L}[v_1:v_{i-1},:],\tilde{\mathbf{X}}). \quad (9.15)$$

Here, we use $\mathbf{L}[v_1:v_{i-1},:]$ to denote the lower-triangular adjacency matrix of the graph that has been generated up to generation step i. The GNN in Equation (9.15) can be instantiated in many ways, but the crucial requirement is that it generates a vector of edge probabilities $\mathbf{L}[v_i,:]$, from which we can sample discrete edge realizations during generation. For example, Liao et al. [2019a] use a variation of the graph attention network (GAT) model (see Chapter 5) to define

---

[2]You et al. [2018] use GRU-style RNNs but in principle LSTMs or other RNN architecture could be employed.

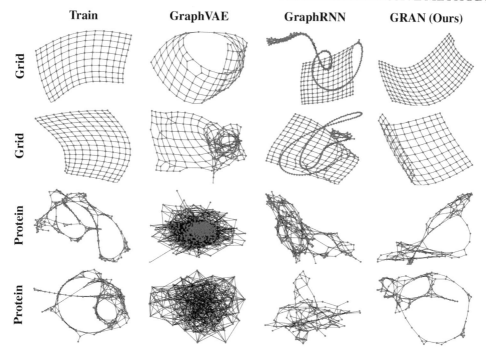

Figure 9.3: Examples of graphs generated by a basic graph-level VAE (Section 9.1), as well as the GraphRNN and GRAN models. Each row corresponds to a different dataset. The first column shows an example of a real graph from the dataset, while the other columns are randomly selected samples of graphs generated by the corresponding model [Liao et al., 2019a].

this GNN. Finally, since there are no node attributes associated with the generated nodes, the input feature matrix $\tilde{\mathbf{X}}$ to the GNN can simply contain randomly sampled vectors (which are useful to distinguish between nodes).

The GRAN model can be trained in an analogous manner as GraphRNN by maximizing the likelihood of training graphs (Equation (9.13)) using teacher forcing. Like the GraphRNN model, we must also specify an ordering over nodes to compute the likelihood on training graphs, and Liao et al. [2019a] provides a detailed discussion on this challenge. Lastly, like the GraphRNN model, we can use GRAN as a generative model after training simply by running the stochastic generation process (e.g., from a fixed initial state), but this model could also be integrated into VAE or GAN-based frameworks.

The key benefit of the GRAN model—compared to GraphRNN—is that it does not need to maintain a long and complex history in a graph-level RNN. Instead, the GRAN model explicitly conditions on the already generated graph using a GNN at each generation step. Liao et al. [2019a] also provide a detailed discussion on how the GRAN model can be optimized to facilitate the generation of large graphs with hundreds of thousands of nodes. For example, one key performance improvement is the idea that multiple nodes can be added simultaneously in a single *block*, rather than adding nodes one at a time. This idea is illustrated in Figure 9.2.

9.4   EVALUATING GRAPH GENERATION

The previous three sections introduced a series of increasingly sophisticated graph generation approaches, based on VAEs, GANs, and autoregressive models. As we introduced these approaches, we hinted at the superiority of some approaches over others. We also provided some examples of generated graphs in Figure 9.3, which hint at the varying capabilities of the different approaches. However, how do we actually quantitatively compare these different models? How can we say that one graph generation approach is better than another? Evaluating generative models is a challenging task, as there is no natural notion of accuracy or error. For example, we could compare reconstruction losses or model likelihoods on held out graphs, but this is complicated by the lack of a uniform likelihood definition across different generation approaches.

In the case of general graph generation, the current practice is to analyze different statistics of the generated graphs, and to compare the distribution of statistics for the generated graphs to a test set [Liao et al., 2019a]. Formally, assume we have set of graph statistics $\mathcal{S} = (s_1, s_2, ..., s_n)$, where each of these statistics $s_{i,\mathcal{G}} : \mathbb{R} \to [0, 1]$ is assumed to define a univariate distribution over $\mathbb{R}$ for a given graph $\mathcal{G}$. For example, for a given graph $\mathcal{G}$, we can compute the degree distribution, the distribution of clustering coefficients, and the distribution of different motifs or graphlets. Given a particular statistic s_i—computed on both a test graph $s_{i,\mathcal{G}_{\text{test}}}$ and a generated graph $s_{i,\mathcal{G}_{\text{gen}}}$—we can compute the distance between the statistic's distribution on the test graph and generated graph using a distributional measure, such as the total variation distance:

$$d(s_{i,\mathcal{G}_{\text{test}}}, s_{i,\mathcal{G}_{\text{gen}}}) = \sup_{x \in \mathbb{R}} |s_{i,\mathcal{G}_{\text{test}}}(x) - s_{i,\mathcal{G}_{\text{gen}}}(x)|. \tag{9.16}$$

To get measure of performance, we can compute the average pairwise distributional distance between a set of generated graphs and graphs in a test set.

Existing works have used this strategy with graph statistics such as degree distributions, graphlet counts, and spectral features, with distributional distances computed using variants of the total variation score and the first Wasserstein distance [Liao et al., 2019b, You et al., 2018].

9.5   MOLECULE GENERATION

All the graph generation approaches we introduced so far are useful for general graph generation. The previous sections did not assume a particular data domain, and our goal was simply to generate realistic graph structures (i.e., adjacency matrices) based on a given training set of graphs. It is worth noting, however, that many works within the general area of graph generation are focused specifically on the task of *molecule generation*.

The goal of molecule generation is to generate molecular graph structures that are both valid (e.g., chemically stable) and ideally have some desirable properties (e.g., medicinal properties or solubility). Unlike the general graph generation problem, research on molecule generation can benefit substantially from domain-specific knowledge for both model design and evaluation strategies. For example, Jin et al. [2018] propose an advanced variant of the graph-level VAE approach (Section 9.1) that leverages knowledge about known molecular motifs. Given the strong dependence on domain-specific knowledge and the unique challenges of molecule generation compared to general graphs, we will not review these approaches in detail here. Nonetheless, it is important to highlight this domain as one of the fastest growing subareas of graph generation.

Conclusion

This book provides a brief (and necessarily incomplete) tour of graph representation learning. Indeed, even as I am writing, there are new and important works arising in this area, and I expect a proper overview of graph representation learning will never be truly complete for many years to come. My hope is that these chapters provide a sufficient foundation and overview for those who are interested in becoming practitioners of these techniques or those who are seeking to explore new methodological frontiers of this area.

My intent is also for these chapters to provide a snapshot of graph representation learning as it stands in what I believe to be a pivotal moment for this nascent area. Recent years have witnessed the formalization of graph representation learning into a genuine and recognizable sub-field within the machine learning community. Spurred by the increased research attention on this topic, GNNs have now become a relatively standard technique; there are now dozens of deep generative models of graphs; and our theoretical understanding of these techniques is solidifying at a rapid pace. However, with this solidification also comes a risk for stagnation, as certain methodologies become ingrained and the focus of research contributions becomes increasingly narrow.

To this end, I will close this book with a brief discussion of two key areas for future work. These are not certainly not the only important areas for inquiry in this field, but they are two areas that I believe hold promise for pushing the fundamentals of graph representation learning forward.

Latent Graph Inference

By and large, the techniques introduced in this book assume that a graph structure is given as an input. The challenge of graph representation learning—as I have presented it—is how to embed or represent such a given input graph in an effective way. However, an equally important and complimentary challenge is the task of *inferring* graphs or relational structure from unstructured or (semi-structured) inputs. This task goes by many names, but I will call it *latent graph inference* here. Latent graph inference is a fundamental challenge for graph representation learning, primarily because it could allow us to use GNN-like methods *even when no input graph is given*. From a technical standpoint, this research direction could potentially build upon the graph generation tools introduced in Part III of this book, ideally combining these tools with the GNN methods introduced in Part II.

Already, there have been promising initial works in this area, such as the Neural Relational Inference (NRI) model proposed by Kipf et al. [2018] and the nearest-neighbor graphs inferred by Wang et al. [2019]. Perhaps the most exciting fact about this research direction is that preliminary findings suggest that latent graph inference might improve model performance *even when we have an input graph*. In my view, building models that can infer latent graph structures beyond the input graph that we are given is a critical direction for pushing forward graph representation learning, which could also open countless new application domains.

Breaking the Bottleneck of Message Passing

Perhaps the single largest topic in this book—in terms of amount of space dedicated—is the neural message passing approach, first introduced in Chapter 5. This message passing formalism—where nodes aggregate messages from neighbors and then update their representations in an iterative fashion—is at the heart of current GNNs and has become the dominant paradigm in graph representation learning.

However, the neural message passing paradigm also has serious drawbacks. As we discussed in Chapter 7, the power of message-passing GNNs are inherently bounded by the WL isomorphism test. Moreover, we know that these message-passing GNNs are theoretically related to relatively simple convolutional filters, which can be formed by polynomials of the (normalized) adjacency matrix. Empirically, researchers have continually found message-passing GNNs to suffer from the problem of over-smoothing, and this issue of over-smoothing can be viewed as a consequence of the neighborhood aggregation operation, which is at the core of current GNNs. Indeed, at their core message-passing GNNs are inherently limited by the aggregate and update message-passing paradigm. This paradigm induces theoretical connections to the WL isomorphism test as well as to simple graph convolutions, but it also induces bounds on the power of these GNNs based on these theoretical constructs.

At a more intuitive level, we can see that the aggregate and update message-passing structure of GNNs inherently induces a tree-structured computation (see, e.g., Figure 5.1). The embedding of each node in a GNN depends on iterative aggregations of neighborhood information, which can be represented as a tree-structured computation graph rooted at that node. Noting that GNNs are restricted to tree-structured computation graph provides yet another view of their limitations, such as their inability to consistently identify cycles and their inability to capture long-range dependencies between the nodes in a graph.

I believe that the core limitations of message-passing GNNs—i.e., being bounded by the WL test, being limited to simple convolutional filters, and being restricted to tree-structured computation graphs—are all, in fact, different facets of a common underlying cause. To push graph representation learning forward, it will be necessary to understand the deeper connections between these theoretical views, and we will need to find new architectures and paradigms that can break these theoretical bottlenecks.

Bibliography

M. Agrawal, M. Zitnik, J. Leskovec, et al. Large-scale analysis of disease pathways in the human interactome. In *PSB*, pages 111–122, 2018. DOI: 10.1101/189787 7

A. Ahmed, N. Shervashidze, S. Narayanamurthy, V. Josifovski, and A. J. Smola. Distributed large-scale natural graph factorization. In *WWW*, 2013. DOI: 10.1145/2488388.2488393 35

R. Albert and L. Barabási. Statistical mechanics of complex networks. *Rev. Mod. Phys.*, 74(1):47, 2002. DOI: 10.1103/revmodphys.74.47 109, 110

J. Lei Ba, J. Ryan Kiros, and Geoffrey E. Hinton. Layer normalization. *ArXiv Preprint ArXiv:1607.06450*, 2016. 76

L. Babai and L. Kucera. Canonical labelling of graphs in linear average time. In *FOCS*, IEEE, 1979. DOI: 10.1109/sfcs.1979.8 96

D. Bahdanau, K. Cho, and Y. Bengio. Neural machine translation by jointly learning to align and translate. In *ICLR*, 2015. 59, 60

P. Barceló, E. Kostylev, M. Monet, J. Pérez, J. Reutter, and J. Silva. The logical expressiveness of graph neural networks. In *ICLR*, 2020. 70

P. Battaglia, et al. Relational inductive biases, deep learning, and graph networks. *ArXiv Preprint ArXiv:1806.01261*, 2018. 69

M. Belkin and P. Niyogi. Laplacian eigenmaps and spectral techniques for embedding and clustering. In *NeurIPS*, 2002. 35

F. Bianchi, D. Grattarola, C. Alippi, and L. Livi. Graph neural networks with convolutional ARMA filters. *ArXiv Preprint ArXiv:1901.01343*, 2019. 87

A. Bojechvski, O. Shchur, D. Zügner, and S. Günnemann. NetGan: Generating graphs via random walks. In *ICML*, 2018. 121

A. Bordes, N. Usunier, A. Garcia-Duran, J. Weston, and O. Yakhnenko. Translating embeddings for modeling multi-relational data. In *NeurIPS*, 2013. 6, 45

K. Borgwardt and H. Kriegel. Shortest-path kernels on graphs. In *ICDM*, 2005. DOI: 10.1109/icdm.2005.132 15

A. Brock, J. Donahue, and K. Simonyan. Large scale GAN training for high fidelity natural image synthesis. In *ICLR*, 2018. 120

J. Bruna, W. Zaremba, A. Szlam, and Y. LeCun. Spectral networks and locally connected networks on graphs. In *ICLR*, 2014. 52, 77, 86

C. Cangea, P. Veličković, N. Jovanović, T. Kipf, and P. Liò. Towards sparse hierarchical graph classifiers. *ArXiv Preprint ArXiv:1811.01287*, 2018. 69

S. Cao, W. Lu, and Q. Xu. GraRep: Learning graph representations with global structural information. In *KDD*, 2015. DOI: 10.1145/2806416.2806512 35

J. Chen, J. Zhu, and L. Song. Stochastic training of graph convolutional networks with variance reduction. In *ICML*, 2018. 76

K. Cho, B. Van Merriënboer, C. Gulcehre, D. Bahdanau, F. Bougares, H. Schwenk, and Y. Bengio. Learning phrase representations using RNN encoder-decoder for statistical machine translation. In *EMNLP*, 2014. DOI: 10.3115/v1/d14-1179 65

A. Clauset, C. Shalizi, and M.E.J. Newman. Power-law distributions in empirical data. *SIAM Rev.*, 51(4):661–703, 2009. DOI: 10.1137/070710111 110

H. Dai, B. Dai, and L. Song. Discriminative embeddings of latent variable models for structured data. In *ICML*, 2016. 52, 77, 90, 93, 94

N. De Cao and T. Kipf. MolGAN: An implicit generative model for small molecular graphs. *ArXiv Preprint ArXiv:1805.11973*, 2018. 121

M. Defferrard, X. Bresson, and P. Vandergheynst. Convolutional neural networks on graphs with fast localized spectral filtering. In *NeurIPS*, 2016. 86

J. Devlin, M. Chang, K. Lee, and K. Toutanova. BERT: Pre-training of deep bidirectional transformers for language understanding. In *NAACL-HLT*, 2018. DOI: 10.18653/v1/N19-1423 61, 74

C. Donnat, M. Zitnik, D. Hallac, and J. Leskovec. Graph wavelets for structural role similarity in complex networks. In *KDD*, 2018. 5

J. Elman. Finding structure in time. *Cog. Sci.*, 14(2):179–211, 1990. DOI: 10.1207/s15516709cog1402_1 55

P. Erdös and A. Rényi. On the evolution of random graphs. *Publ. Math. Inst. Hung. Acad. Sci.*, 5(1):17–60, 1960. 108

H. Gao and S. Ji. Graph U-Nets. In *ICML*, 2019. 69

J. Gilmer, S. Schoenholz, P. Riley, O. Vinyals, and G. Dahl. Neural message passing for quantum chemistry. In *ICML*, 2017. 7, 52

I. Goodfellow, J. Pouget-Abadie, M. Mirza, B. Xu, D. Warde-Farley, S. Ozair, A. Courville, and Y. Bengio. Generative adversarial nets. In *NeurIPS*, 2014. 120

I. Goodfellow, Y. Bengio, and A. Courville. *Deep Learning*. MIT Press, 2016. xv, 73, 114

L. Grafakos. *Classical and Modern Fourier Analysis*. Prentice Hall, 2004. 79

A. Grover and J. Leskovec. node2vec: Scalable feature learning for networks. In *KDD*, 2016. DOI: 10.1145/2939672.2939754 37

A. Grover, A. Zweig, and S. Ermon. Graphite: Iterative generative modeling of graphs. In *ICML*, 2019. 117

W. Hamilton, R. Ying, and J. Leskovec. Representation learning on graphs: Methods and applications. *IEEE Data Eng. Bull.*, 2017a. 31, 34, 64

W. L. Hamilton, R. Ying, and J. Leskovec. Inductive representation learning on large graphs. In *NeurIPS*, 2017b. 4, 39, 52, 58, 62, 75, 77

K. He, X. Zhang, S. Ren, and J. Sun. Deep residual learning for image recognition. In *CVPR*, 2016. DOI: 10.1109/cvpr.2016.90 64

M. Heusel, H. Ramsauer, T. Unterthiner, B. Nessler, and S. Hochreiter. GANs trained by a two time-scale update rule converge to a local Nash equilibrium. In *NeurIPS*, 2017. 120

S. Hochreiter and J. Schmidhuber. Long short-term memory. *Neur. Comput.*, 9(8):1735–1780, 1997. DOI: 10.1162/neco.1997.9.8.1735 59

P. Hoff, A. E. Raftery, and M. S. Handcock. Latent space approaches to social network analysis. *JASA*, 97(460):1090–1098, 2002. DOI: 10.1198/016214502388618906 31

S. Hoory, N. Linial, and A. Wigderson. Expander graphs and their applications. *Bull. Am. Math. Soc.*, 43(4):439–561, 2006. DOI: 10.1090/s0273-0979-06-01126-8 63

W. Hu, B. Liu, J. Gomes, M. Zitnik, P. Liang, V. Pande, and J. Leskovec. Strategies for pre-training graph neural networks. In *ICLR*, 2019. 74

Matthew O. Jackson. *Social and Economic Networks*. Princeton University Press, 2010. 10

G. Ji, S. He, L. Xu, K. Liu, and J. Zhao. Knowledge graph embedding via dynamic mapping matrix. In *ACL*, 2015. DOI: 10.3115/v1/p15-1067 46

W. Jin, R. Barzilay, and T. Jaakkola. Junction tree variational autoencoder for molecular graph generation. In *ICML*, 2018. 127

H. Kashima, K. Tsuda, and A. Inokuchi. Marginalized kernels between labeled graphs. In *ICML*, 2003. 15

Y. Katznelson. *An Introduction to Harmonic Analysis*. Cambridge University Press, 2004. 79

D. Kingma and M. Welling. Auto-encoding variational Bayes. *ArXiv Preprint ArXiv:1312.6114*, 2013. 114

T. Kipf, E. Fetaya, K. Wang, M. Welling, and R. Zemel. Neural relational inference for interacting systems. In *ICML*, 2018. 129

T. N. Kipf and M. Welling. Semi-supervised classification with graph convolutional networks. In *ICLR*, 2016a. 4, 57, 72, 87, 88

T. N. Kipf and M. Welling. Variational graph auto-encoders. In *NeurIPS Workshop on Bayesian Deep Learning*, 2016b. 116, 117

J. Klicpera, A. Bojchevski, and S. Günnemann. Predict then propagate: Graph neural networks meet personalized PageRank. In *ICLR*, 2019. 89

N. Kriege, F. Johansson, and C. Morris. A survey on graph kernels. *Appl. Netw. Sci.*, 5(1):1–42, 2020. DOI: 10.1007/s41109-019-0195-3 14

J. B. Kruskal. Multidimensional scaling by optimizing goodness of fit to a nonmetric hypothesis. *Psychometrika*, 29(1):1–27, 1964. DOI: 10.1007/bf02289565 35

E. Leicht, P. Holme, and M. E. J. Newman. Vertex similarity in networks. *Phys. Rev. E*, 73(2):026120, 2006. DOI: 10.1103/physreve.73.026120 19, 20

J. Leskovec, A. Rajaraman, and J. Ullman. *Mining of Massive Data Sets*. Cambridge University Press, 2020. 21

R. Levie, F. Monti, X. Bresson, and M. Bronstein. Cayleynets: Graph convolutional neural networks with complex rational spectral filters. *IEEE Trans. Signal Process*, 67(1):97–109, 2018. DOI: 10.1109/tsp.2018.2879624 87

Y. Li, D. Tarlow, M. Brockschmidt, and R. Zemel. Gated graph sequence neural networks. In *ICLR*, 2015. 65, 76

Y. Li, O. Vinyals, C. Dyer, R. Pascanu, and P. Battaglia. Learning deep generative models of graphs. In *ICML*, 2018. 122

Y. Li, C. Gu, T. Dullien, O. Vinyals, and P. Kohli. Graph matching networks for learning the similarity of graph structured objects. In *ICML*, 2019. 7

R. Liao, Y. Li, Y. Song, S. Wang, W. L. Hamilton, D. Duvenaud, R. Urtasun, and R. Zemel. Efficient graph generation with graph recurrent attention networks. In *NeurIPS*, 2019a. 119, 122, 124, 125, 126

R. Liao, Z. Zhao, R. Urtasun, and R. Zemel. LanczosNet: Multi-scale deep graph convolutional networks. In *ICLR*, 2019b. 87, 126

L. Lü and T. Zhou. Link prediction in complex networks: A survey. *Physica A*, 390(6):1150–1170, 2011. DOI: 10.1016/j.physa.2010.11.027 6, 17, 20

D. Marcheggiani and I. Titov. Encoding sentences with graph convolutional networks for semantic role labeling. In *EMNLP*, 2017. DOI: 10.18653/v1/d17-1159 67

H. Maron, H. Ben-Hamu, H. Serviansky, and Y. Lipman. Provably powerful graph networks. In *NeurIPS*, 2019. 101, 102, 103

J. Mason and D. Handscomb. *Chebyshev Polynomials*. Chapman and Hall, 2002. DOI: 10.1201/9781420036114 87

M. McPherson, L. Smith-Lovin, and J. Cook. Birds of a feather: Homophily in social networks. *Annu. Rev. Sociol.*, 27(1):415–444, 2001. DOI: 10.1146/annurev.soc.27.1.415 5

C. Merkwirth and L. Lengauer. Automatic generation of complementary descriptors with molecular graph, In *J. Chem. Inf. Model*, 45(5):1159–1168, 2005. 54

L. Mescheder, A. Geiger, and S. Nowozin. Which training methods for GANs do actually converge? In *ICML*, 2018. 120

C. D. Meyer. *Matrix Analysis and Applied Linear Algebra*, vol. 71. SIAM, 2000. DOI: 10.1137/1.9780898719512 11

C. Morris, M. Ritzert, M. Fey, W. L. Hamilton, J. Lenssen, G. Rattan, and M. Grohe. Weisfeiler and Leman go neural: Higher-order graph neural networks. In *AAAI*, 2019. DOI: 10.1609/aaai.v33i01.33014602 73, 97, 98, 101

R. Murphy, B. Srinivasan, V. Rao, and B. Ribeiro. Janossy pooling: Learning deep permutation-invariant functions for variable-size inputs. In *ICLR*, 2018. 58, 59

R. Murphy, B. Srinivasan, V. Rao, and B. Ribeiro. Relational pooling for graph representations. In *ICML*, 2019. 99, 100

M. Newman. Mathematics of networks. *The New Palgrave Dictionary of Economics*, pages 1–8, 2016. 12

M. Newman. *Networks*. Oxford University Press, 2018. 1, 12, 13, 108, 109

D. Nguyen, K. Sirts, L. Qu, and M. Johnson. STranse: A novel embedding model of entities and relationships in knowledge bases. *ArXiv Preprint ArXiv:1606.08140*, 2016. DOI: 10.18653/v1/n16-1054 46

M. Nickel, V. Tresp, and H. Kriegel. A three-way model for collective learning on multi-relational data. In *ICML*, 2011. 42

M. Nickel, K. Murphy, V. Tresp, and E. Gabrilovich. A review of relational machine learning for knowledge graphs. *Proc. IEEE*, 104(1):11–33, 2016. DOI: 10.1109/jproc.2015.2483592 6, 41

A. Oppenheim, R. Schafer, and J. Buck. *Discrete-Time Signal Processing*. Prentice Hall, 1999. 79

A. Ortega, P. Frossard, J. Kovačević, J. Moura, and P. Vandergheynst. Graph signal processing: Overview, challenges, and applications. *Proc. IEEE*, 106(5):808–828, 2018. DOI: 10.1109/jproc.2018.2820126 27, 81, 82

M. Ou, P. Cui, J. Pei, Z. Zhang, and W. Zhu. Asymmetric transitivity preserving graph embedding. In *KDD*, 2016. DOI: 10.1145/2939672.2939751 35

J. Padgett and C. Ansell. Robust action and the rise of the medici, 1400–1434. *Am. J. Sociol.*, 98(6):1259–1319, 1993. DOI: 10.1086/230190 10

L. Page, S. Brin, R. Motwani, and T. Winograd. The PageRank citation ranking: Bringing order to the web. Technical report, Stanford InfoLab, 1999. 21

S. Pandit, D. Chau, S. Wang, and C. Faloutsos. NetProbe: A fast and scalable system for fraud detection in online auction networks. In *WWW*, 2007. DOI: 10.1145/1242572.1242600 7

B. Perozzi, R. Al-Rfou, and S. Skiena. Deepwalk: Online learning of social representations. In *KDD*, 2014. DOI: 10.1145/2623330.2623732 18, 37

B. Perozzi, V. Kulkarni, and S. Skiena. Walklets: Multiscale graph embeddings for interpretable network classification. *ArXiv Preprint ArXiv:1605.02115*, 2016. 38

T. Pham, T. Tran, D. Phung, and S. Venkatesh. Column networks for collective classification. In *AAAI*, 2017. 64

C. R. Qi, H. Su, K. Mo, and L. J. Guibas. Pointnet: Deep learning on point sets for 3d classification and segmentation. In *CVPR*, 2017. DOI: 10.1109/cvpr.2017.16 58

J. Qiu, Y. Dong, H. Ma, J. Li, K. Wang, and J. Tang. Network embedding as matrix factorization: Unifying deepwalk, line, pte, and node2vec. In *KDD*, 2018. DOI: 10.1145/3159652.3159706 38, 39

M. Qu, Y. Bengio, and J. Tang. GMNN: Graph Markov neural networks. In *ICML*, 2019. 94

L. Rabiner and B. Gold. *Theory and Application of Digital Signal Processing*. Prentice-Hall, 1975. DOI: 10.1109/tsmc.1978.4309918 79

L. F. R. Ribeiro, P. H. P. Saverese, and D. R. Figueiredo. struc2vec: Learning node representations from structural identity. In *KDD*, 2017. DOI: 10.1145/3097983.3098061 38

H. Robbins and S. Monro. A stochastic approximation method. *Ann. Math. Stat.*, pages 400–407, 1951. 34

D. Rumelhart, G. Hinton, and R. Williams. Learning representations by back-propagating errors. *Nature*, 323(6088):533–536, 1986. DOI: 10.1038/323533a0 71

F. Scarselli, M. Gori, A. C. Tsoi, M. Hagenbuchner, and G. Monfardini. The graph neural network model. *IEEE Trans. Neural Netw. Learn. Syst.*, 20(1):61–80, 2009. DOI: 10.1109/tnn.2008.2005605 54

M. Schlichtkrull, T. N. Kipf, P. Bloem, R. van den Berg, I. Titov, and M. Welling. Modeling relational data with graph convolutional networks. In *European Semantic Web Conference*, 2017. DOI: 10.1007/978-3-319-93417-4_38 41, 66, 73, 76

D. Selsam, M. Lamm, B. Bünz, P. Liang, L. de Moura, and D. Dill. Learning a SAT solver from single-bit supervision. In *ICLR*, 2019. 65, 76

N. Shervashidze and K. Borgwardt. Fast subtree kernels on graphs. In *NeurIPS*, 2009. 15

N. Shervashidze, P. Schweitzer, E. Leeuwen, K. Mehlhorn, and K. Borgwardt. Weisfeiler-Lehman graph kernels. *JMLR*, 12:2539–2561, 2011. 14, 15

M. Simonovsky and N. Komodakis. GraphVAE: Towards generation of small graphs using variational autoencoders. In *International Conference on Artificial Neural Networks*, 2018. DOI: 10.1007/978-3-030-01418-6_41 117, 118, 119

K. Sinha, S. Sodhani, J. Dong, J. Pineau, and W. Hamilton. CLUTRR: A diagnostic benchmark for inductive reasoning from text. In *EMNLP*, 2019. DOI: 10.18653/v1/d19-1458 67

A. Smola, A. Gretton, L. Song, and B. Schölkopf. A Hilbert space embedding for distributions. In *COLT*, 2007. DOI: 10.1007/978-3-540-75225-7_5 90

N. Srivastava, G. Hinton, A. Krizhevsky, I. Sutskever, and R. Salakhutdinov. Dropout: A simple way to prevent neural networks from overfitting. *JMLR*, 15(1):1929–1958, 2014. 76

M. Stoer and F. Wagner. A simple min-cut algorithm. *J. ACM*, 44(4):585–591, 1997. DOI: 10.1145/263867.263872 24

F. Sun, J. Hoffmann, and J. Tang. Infograph: Unsupervised and semi-supervised graph-level representation learning via mutual information maximization. In *ICLR*, 2020. 74

Z. Sun, Z. Deng, J. Nie, and J. Tang. RotatE: Knowledge graph embedding by relational rotation in complex space. In *ICLR*, 2019. 44, 47, 48

J. Tang, M. Qu, M. Wang, M. Zhang, J. Yan, and Q. Mei. LINE: Large-scale information network embedding. In *WWW*, 2015. DOI: 10.1145/2736277.2741093 38

K. Teru, E. Denis, and W. L. Hamilton. Inductive relation prediction on knowledge graphs. In *ICML*, 2020. 6, 67, 76

T. Trouillon, J. Welbl, S. Riedel, É. Gaussier, and G. Bouchard. Complex embeddings for simple link prediction. In *ICML*, 2016. 46

A. Vaswani, N. Shazeer, N. Parmar, J. Uszkoreit, L. Jones, A. Gomez, L. Kaiser, and I. Polosukhin. Attention is all you need. In *NeurIPS*, 2017. 60, 61

P. Veličković, G. Cucurull, A. Casanova, A. Romero, P. Lio, and Y. Bengio. Graph attention networks. In *ICLR*, 2018. 60, 76

P. Veličković, W. Fedus, W. L. Hamilton, P. Liò, Y. Bengio, and R. D. Hjelm. Deep graph infomax. In *ICLR*, 2019. 74

O. Vinyals, S. Bengio, and M. Kudlur. Order matters: Sequence to sequence for sets. In *ICLR*, 2015. 68

S. V. N. Vishwanathan, N. N. Schraudolph, R. Kondor, and K. M. Borgwardt. Graph kernels. *JMLR*, 11:1201–1242, 2010. DOI: 10.1142/9789812772435_0002 14

U. Von Luxburg. A tutorial on spectral clustering. *Stat. Comput.*, 17(4):395–416, 2007. DOI: 10.1007/s11222-007-9033-z 26, 27

M. Wainwright and M. Jordan. Graphical models, exponential families, and variational inference. *Found. Trends Mach. Learn.*, 1(1–2):1–305, 2008. DOI: 10.1561/2200000001 90, 91, 115

Y. Wang, Y. Sun, Z. Liu, S. Sarma, M. Bronstein, and J. Solomon. Dynamic graph CNN for learning on point clouds. *ACM TOG*, 38(5):1–12, 2019. DOI: 10.1145/3326362 129

Z. Wang, J. Zhang, J. Feng, and Z. Chen. Knowledge graph embedding by translating on hyperplanes. In *AAAI*, 2014. 46

Duncan J. Watts and Steven H. Strogatz. Collective dynamics of "small-world" networks. *Nature*, 393(6684):440–442, 1998. DOI: 10.1038/30918 12, 13

B. Weisfeiler and A. Lehman. A reduction of a graph to a canonical form and an algebra arising during this reduction. *Nauchno-Technicheskaya Informatsia*, 2(9):12–16, 1968. 14

R. Williams and D. Zipser. A learning algorithm for continually running fully recurrent neural networks. *Neural Comput.*, 1(2):270–280, 1989. DOI: 10.1162/neco.1989.1.2.270 123

F. Wu, T. Zhang, C. Souza, A. Fifty, T. Yu, and K. Weinberger. Simplifying graph convolutional networks. In *ICML*, 2019. 89

K. Xu, C. Li, Y. Tian, T. Sonobe, K. Kawarabayashi, and S. Jegelka. Representation learning on graphs with jumping knowledge networks. In *ICML*, 2018. 62, 63, 65

K. Xu, W. Hu, J. Leskovec, and S. Jegelka. How powerful are graph neural networks? In *ICLR*, 2019. 97, 98

B. Yang, W. Yih, X. He, J. Gao, and L. Deng. Embedding entities and relations for learning and inference in knowledge bases. In *ICLR*. 46

Z. Yang, W. Cohen, and R. Salakhutdinov. Revisiting semi-supervised learning with graph embeddings. In *ICML*, 2016. 5

Z. Yang, Z. Dai, Y. Yang, J. Carbonell, R. Salakhutdinov, and Q. Le. XLnet: Generalized autoregressive pretraining for language understanding. In *NeurIPS*, 2019. 61

R. Ying, R. He, K. Chen, P. Eksombatchai, W. L. Hamilton, and J. Leskovec. Graph convolutional neural networks for web-scale recommender systems. In *KDD*, 2018a. DOI: 10.1145/3219819.3219890 6, 73, 76

R. Ying, J. You, C. Morris, X. Ren, W. Hamilton, and J. Leskovec. Hierarchical graph representation learning with differentiable pooling. In *NeurIPS*, 2018b. 69

J. You, R. Ying, X. Ren, W. L. Hamilton, and J. Leskovec. GraphRNN: Generating realistic graphs with deep auto-regressive models. In *ICML*, 2018. 122, 123, 124, 126

W. Zachary. An information flow model for conflict and fission in small groups. *J. Anthropol. Res.*, 33(4):452–473, 1977. DOI: 10.1086/jar.33.4.3629752 2

M. Zaheer, S. Kottur, S. Ravanbakhsh, B. Poczos, R. Salakhutdinov, and A. Smola. Deep sets. In *NeurIPS*, 2017. 58

Y. Zhang, X. Chen, Y. Yang, A. Ramamurthy, B. Li, Y. Qi, and L. Song. Can graph neural networks help logic reasoning? In *ICLR*, 2020. 94

D. Zhou, O. Bousquet, T. Lal, J. Weston, and B. Schölkopf. Learning with local and global consistency. In *NeurIPS*, 2004. 5

M. Zitnik, M. Agrawal, and J. Leskovec. Modeling polypharmacy side effects with graph convolutional networks. *Bioinformatics*, 34(13):457–i466, 2018. DOI: 10.1093/bioinformatics/bty294 6, 67

Author's Biography

WILLIAM L. HAMILTON

William L. Hamilton is an Assistant Professor of Computer Science at McGill University and a Canada CIFAR Chair in AI. His research focuses on graph representation learning as well as applications in computational social science and biology. In recent years, he has published more than 20 papers on graph representation learning at top-tier venues across machine learning and network science, as well as co-organized several large workshops and tutorials on the topic. William's work has been recognized by several awards, including the 2018 Arthur L. Samuel Thesis Award for the best doctoral thesis in the Computer Science department at Stanford University and the 2017 Cozzarelli Best Paper Award from the *Proceedings of the National Academy of Sciences.*

Printed in the United States
by Baker & Taylor Publisher Services